The Essentials of Probability

The Essentials of Probability

Richard Durrett

Cornell University

Duxbury Press
An Imprint of Wadsworth Publishing Company
Belmont, California

Production Editor: *Gary Mcdonald*

Managing Designer: *Cloyce Wall*

Print Buyer: *Diana Spence*

Art Editor: *Donna Kalal*

Copy Editor: *Judith Abrahms*

Technical Illustrator: *Teresa Roberts*

Cover Designer: *Cloyce Wall*

Printer: *Maple-Vail Book Manufacturing Group*

International Thomson Publishing
The trademark ITP is used under license.

*This book is printed on
acid-free recycled paper.*

Duxbury Press
An Imprint of Wadsworth Publishing Company
A Division of Wadsworth, Inc.

Printed in the United States of America
1 2 3 4 5 6 7 8 9 10—98 97 96 95 94

Library of Congress Cataloging-in-Publication Data
Durrett, Richard, 1951–
    The essentials of probability / Richard Durrett.
       p.   cm.
    Includes index.
    ISBN 0-534-19230-0
    1. Probabilities.   I. Title.
QA273.D864   1994
519.2—dc20   93-29042

Preface

*Probability is the most important concept in modern science,
especially as nobody has the slightest notion what it means.*

BERTRAND RUSSELL

This book is designed for Math 471 at Cornell University, a one-semester course in probability that is taken by math majors and students from other departments, many of whom continue and take the second-semester course in statistics. As the title and chapter headings indicate, we will focus on the essentials of probability theory: (1) Coins, Dice and Cards (so-called Combinatorial Probability), (2) Conditional Probability, (3) Distributions, (4) Expected Value, and (5) Limit Theorems. It is difficult to imagine someone covering the entire book in one semester, so we have starred some sections that are not needed in the main quest of this book: the laws of large numbers and the central limit theorem.

We take our literary style from Joe Friday's famous line on Dragnet, "Just the facts, ma'am." Theoretical results are described briefly and are then illustrated by a number of examples. Our philosophy is that the goal of probability theory is solving problems, so we have included more than 650 exercises. Some of these are in the text itself while others are at the end of the section. The ones in the text are meant to (a) illustrate the concepts just introduced or (b) give extensions of the results presented. All students should do the exercises of type (a) as they go along and should check their answers with the ones given in the back of the book. The end-of-the-section exercises are designed to be "homework problems." In keeping with tradition, only answers to even-numbered end-of-the-section exercises are given and they are given in a form that allows the student to check her answer but not to infer the solution method from it.

Several people have read alpha and beta versions of this book and pointed out numerous typos and errors. In this connection I would especially like to thank Semyon Krugyak and Scott Arouh, two Cornell undergraduates who read the book. In addition, I benefited tremendously from hired critics L. Gray, D. Griffeath, and G. Lawler, and from six anonymous reviewers. Closer to home, I would like to thank my two sons, Gregory and David, now 4 and 6, for their help in writing the book. The hours we spent at home with Duke Nukum and Commander Keen (PC games from Apogee Software Products), and at the mall with the Simpsons and X-men, provided a necessary break from the writing process.

Turning to those to whom I am bound by various contracts, I would like to give my traditional thanks to my wife Susan for her patience and understanding, each of which has been tested more than once in the last twenty, some odd, months. I would like to say a warm "Thank you" to John Kimmel who brought my first three books into the world and signed the contract for this one, before the *Invasion of the Bean Counters* sent him to Chapman and Hall and left me in the capable hands of Alex Kugushev. Last and perhaps least, this book could not have been written without (are you listening, Tipper Gore?) some loud music in the background. If you put your ear close to the book you may be able to hear music by Metallica, Nirvana, Pearl Jam, or Van Halen, but you will certainly look silly doing so.

At about the time I was finishing version alpha of this manuscript in January 1992, my colleague and friend Frank Spitzer died. He was a kind and gentle man and I would like to dedicate this book to him.

Rick Durrett

Contents

4. Expected Value

5. Limit Theorems

1 Coins, Dice, and Cards

1.1. Experiments and Events

Probability has its roots in the analysis of gambling games, so that is where we will start. We begin with a vague but useful definition. (Here and in what follows, **boldface** indicates a word or phrase that is being defined or explained.) The term **experiment** is used to refer to any process whose outcome is not known in advance. Three simple experiments are: flip a coin, roll a die, and pick a card from a deck. These examples are probably somewhat familiar, but we will define them precisely after we give the next definition. The **sample space** associated with an experiment is the set of all possible outcomes. The sample space is usually denoted by Ω, the capital Greek letter Omega.

Example 1.1. Flip a coin and let it fall on the ground. Most coins have a person's face on one side (Heads) and something else on the other (Tails). The outcome of the experiment is the name of the side that is showing, which we indicate by writing the first letter of the word, so $\Omega = \{H, T\}$.

Example 1.2. A die is a cube with the numbers 1, 2, 3, 4, 5, and 6 on its six sides, with each number usually represented by that number of dots. The outcome of throwing a die is the number on the top when it comes to rest, so $\Omega = \{1, 2, 3, 4, 5, 6\}$.

Example 1.3. A deck of 52 cards consists of 13 types of cards: Ace, King, Queen, Jack, 10, 9, 8, 7, 6, 5, 4, 3, and 2, each of which comes in four varieties – spade ($\spadesuit$), heart ($\heartsuit$), diamond ($\diamondsuit$), and club ($\clubsuit$) – for a total of 52 outcomes:

$$\Omega = \{A\spadesuit, K\spadesuit, Q\spadesuit, J\spadesuit, 10\spadesuit, 9\spadesuit, 8\spadesuit, 7\spadesuit, 6\spadesuit, 5\spadesuit, 4\spadesuit, 3\spadesuit, 2\spadesuit$$
$$A\heartsuit, K\heartsuit, Q\heartsuit, J\heartsuit, 10\heartsuit, 9\heartsuit, 8\heartsuit, 7\heartsuit, 6\heartsuit, 5\heartsuit, 4\heartsuit, 3\heartsuit, 2\heartsuit$$
$$A\diamondsuit, K\diamondsuit, Q\diamondsuit, J\diamondsuit, 10\diamondsuit, 9\diamondsuit, 8\diamondsuit, 7\diamondsuit, 6\diamondsuit, 5\diamondsuit, 4\diamondsuit, 3\diamondsuit, 2\diamondsuit$$
$$A\clubsuit, K\clubsuit, Q\clubsuit, J\clubsuit, 10\clubsuit, 9\clubsuit, 8\clubsuit, 7\clubsuit, 6\clubsuit, 5\clubsuit, 4\clubsuit, 3\clubsuit, 2\clubsuit\}$$

(Here and in what follows A, K, Q, and J are abbreviations for Ace, King, Queen, and Jack.)

The goal of probability theory is to compute the probability of various events of interest. Intuitively, an event is a statement about the outcome of an experiment. The formal definition is: An **event** is a subset of the sample space. For example, "The coin shows Heads" $= \{H\}$, or "The die shows an even number" $= \{2, 4, 6\}$. To create some more interesting examples we need some more elaborate experiments.

Example 1.4. Roll two dice that, for convenience, we assume are red and green. We write the outcomes of this experiment as (m, n), where m is the number on the red die and n is the number on the green die. To visualize the set of outcomes it is useful to make a little table:

$$\begin{array}{cccccc}
(1,1) & (2,1) & (3,1) & (4,1) & (5,1) & (6,1) \\
(1,2) & (2,2) & (3,2) & (4,2) & (5,2) & (6,2) \\
(1,3) & (2,3) & (3,3) & (4,3) & (5,3) & (6,3) \\
(1,4) & (2,4) & (3,4) & (4,4) & (5,4) & (6,4) \\
(1,5) & (2,5) & (3,5) & (4,5) & (5,5) & (6,5) \\
(1,6) & (2,6) & (3,6) & (4,6) & (5,6) & (6,6)
\end{array}$$

There are $36 = 6 \cdot 6$ outcomes since there are 6 rows and 6 columns, or, using reasoning that generalizes more easily to three dice, there are 6 possible numbers to write in the first slot and for each number written in the first slot there are 6 possibilities for the second. An example of an event concerning two dice is "The sum is 8" $= \{(2, 6), (3, 5), (4, 4), (5, 3), (6, 2)\}$.

EXERCISE 1.1. How many outcomes are in the events (a) "The sum is 9," (b) "The sum is 10"?

Example 1.5. Flip three coins. Extending the reasoning from the last example we see there are $8 = 2 \cdot 2 \cdot 2$ possible outcomes:

$$\begin{array}{cccc}
 & (H, H, T) & (T, T, H) & \\
(H, H, H) & (H, T, H) & (T, H, T) & (T, T, T) \\
 & (T, H, H) & (H, T, T) &
\end{array}$$

An example of an event concerning three coins is "We get two Heads" $= \{(H, H, T), (H, T, H), (T, H, H)\}$.

EXERCISE 1.2. How many outcomes are there if we roll three dice?

EXERCISE 1.3. (a) How many outcomes are there if we flip four coins? How many outcomes are in (b) "We get one Head," (c) "We get two Heads"?

In working with events we will need some concepts from set theory, which we now review. In the pictures below, A is a triangle, B is a rectangle with two broken corners, and Ω, when it appears, is a rectangle. We say that A is a **subset** of B, and write $A \subset B$, if every outcome in A is also in B.

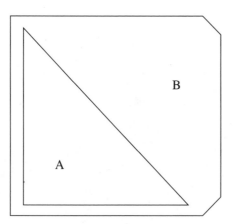

The **union** of A and B, $A \cup B$, is the set of outcomes that are in A or B.

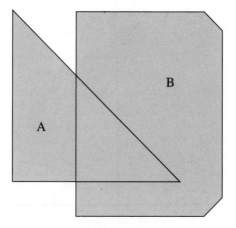

The **intersection** of A and B, $A \cap B$, is the set of outcomes that are in both A and B.

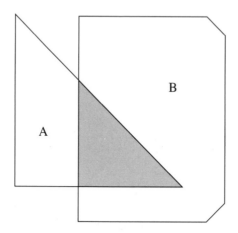

The **empty set**, $\emptyset$, is the event with no outcomes. Two events are **disjoint** if they have no outcomes in common, or $A \cap B = \emptyset$.

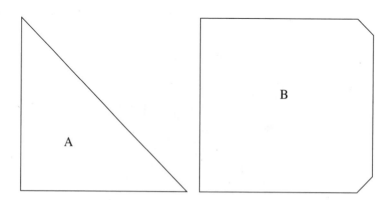

The **complement** of A, A^c, is the set of outcomes not in A.

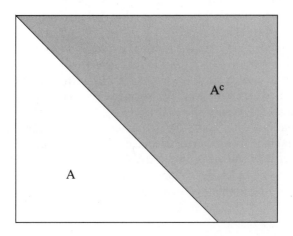

The **difference** $B - A$ is the set of outcomes in B but not in A, or simply $B \cap A^c$.

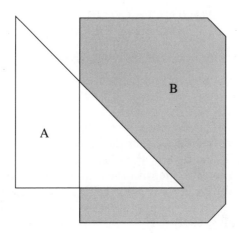

EXERCISE 1.4. Suppose $\Omega = \{1, 2, 3, 4, 5, 6\}$, $A = \{1, 2\}$, and $B = \{2, 3, 4\}$. Compute $A \cup B$, $A \cap B$, A^c, and $B - A$.

A finite sequence $A_1, A_2, \ldots A_n$ or an infinite sequence $A_1, A_2, \ldots$ of events is said to be **disjoint** if $A_i \cap A_j = \emptyset$ whenever $i \neq j$. The definitions of union and intersection extend in the obvious way to sequences of events. The **union of** $A_1, A_2, \ldots A_n$ is the set of outcomes that belong to at least one of these events and is written as either $A_1 \cup A_2 \ldots \cup A_n$ or $\cup_{i=1}^n A_i$. The **union of the infinite sequence** $A_1, A_2, \ldots$ is the set of outcomes that are in at least one of the events and is written as $\cup_{i=1}^\infty A_i$. In some cases when the exact index set is unimportant or is clear from the context we will write simply $\cup_i A_i$. Similarly, the **intersection** of a finite or infinite sequence of events is the set of outcomes that belong to all the events and is denoted by $\cap_i A_i$.

In addition to events, we will be interested in random variables. Intuitively, a random variable is a number determined by the outcome of an experiment. Formally, a **random variable** is a real-valued function defined on the sample space. Concrete examples of random variables are the sum of the two dice in Example 1.4 and the number of Heads in Example 1.5.

EXERCISES

1.5. Let A be the event that a person is male, B that the person is under 30, and C that the person speaks French. Describe in symbols (a) a male over 30, (b) a female who is under 30 and speaks French, (c) a male who either is under 30 or speaks French.

1.6. Prove the **distributive laws**

$$A \cap (B \cup C) = (A \cap B) \cup (A \cap C)$$
$$A \cup (B \cap C) = (A \cup B) \cap (A \cup C)$$

(Here the name comes from the analogy with $a \cdot (b + c) = (a \cdot b) + (a \cdot c)$.)

1.7. Show that $(\cap_i A_i)^c = \cup_i A_i^c$ and $(\cup_i A_i)^c = \cap_i A_i^c$.

1.2. Probabilities

In the previous section we learned that the set of all outcomes of an experiment is the sample space Ω, and that an event is a subset of the sample space. These objects will begin to come to life with the introduction of the the third crucial ingredient. **A probability** is a way of assigning numbers to events that satisfies

(i) For any event A, $0 \leq P(A) \leq 1$.

(ii) If Ω is the sample space then $P(\Omega) = 1$.

(iii) For a finite or infinite sequence of disjoint events

$$P(\cup_i A_i) = \sum_i P(A_i)$$

In words, the probability of union of disjoint events is the sum of the probabilities of the sets. Here we are using the convention announced at the end of the last section. When the exact index set is unimportant we will write $\cup_i$ or $\sum_i$. For a finite sequence of events, (iii) says

$$P(A_1 \cup A_2 \cdots \cup A_n) = P(A_1) + P(A_2) + \cdots + P(A_n)$$

or $P(\cup_{i=1}^n A_i) = \sum_{i=1}^n P(A_i)$. For an infinite sequence of events, (iii) says

$$P(\cup_{i=1}^\infty A_i) = \sum_{i=1}^\infty P(A_i)$$

These assumptions are motivated by the **frequency interpretation of probability**, which states that if we repeat an experiment a large number of times then the fraction of times the event A occurs will be close to $P(A)$. To be precise, if we let $N(A, n)$ be the number of times A occurs in the first n trials then

$$P(A) = \lim_{n \to \infty} N(A, n)/n$$

Given this interpretation, (i) and (ii) are clear. The fraction of times that A occurs must be between 0 and 1, and if Ω has been defined properly (recall that it is the set of ALL possible outcomes), the fraction of times something in Ω happens is 1. To explain (iii), note that if the A_i are disjoint then

$$N(\cup_i A_i, n) = \sum_i N(A_i, n)$$

since when $\cup_i A_i$ occurs exactly one of the A_i does. Dividing by n and letting $n \to \infty$, we arrive at (iii). The previous sentence is not meant to prove (iii), just to explain why it is a reasonable assumption.

To explain what probabilities look like, we consider

Example 2.1. Suppose $\Omega = \{1, 2, 3\}$ and

$$P(\{1\}) = p_1 \qquad P(\{2\}) = p_2 \qquad P(\{3\}) = p_3$$

where $p_1, p_2, p_3 \geq 0$. It follows from (iii) that

$$P(\{1, 2\}) = P(\{1\}) + P(\{2\}) = p_1 + p_2$$
$$P(\{1, 3\}) = P(\{1\}) + P(\{3\}) = p_1 + p_3$$
$$P(\{2, 3\}) = P(\{2\}) + P(\{3\}) = p_2 + p_3$$
$$P(\Omega) = P(\{1\}) + P(\{2\}) + P(\{3\}) = p_1 + p_2 + p_3$$

so the last equation and (ii) imply that $p_1 + p_2 + p_3 = 1$. Conversely, if we let $P(\{i\}) = p_i$ where $p_1, p_2, p_3 \geq 0$ and $p_1 + p_2 + p_3 = 1$ then the equations above define a probability. Generalizing from this example:

Finite Sample Spaces. Suppose $\Omega = \{1, 2, \ldots k\}$. In this case, the probability is determined by giving the probabilities $p_j = P(\{j\})$ with $p_1, \ldots, p_k \geq 0$ and $p_1 + \cdots + p_k = 1$ since (iii) allows us to compute the probabilities of larger sets. For example, $P(\{1, 4\}) = P(\{1\}) + P(\{4\})$ and $P(\{2, 3, 5\}) = P(\{2\}) + P(\{3\}) + P(\{5\})$. In general,

$$P(A) = \sum_{j \in A} P(\{j\})$$

or simply "Compute $P(A)$ by adding the probabilities of all the outcomes j in A." We leave it to the reader to check that this definition satisfies (iii) and hence defines a probability.

An important special case (which includes coins, dice, and cards) occurs when all the outcomes have the same probability, c. In this case, if we use $|S|$ to denote the number of elements in a set S then

$$P(A) = \sum_{j \in A} P(\{j\}) = c|A|$$

Taking $A = \Omega$ in the last equation shows that we must pick $c = 1/|\Omega|$ to make $P(\Omega) = 1$. If we do this then $P(A) = |A|/|\Omega|$, i.e., the fraction of outcomes that lie in A. For a concrete example, consider

Example 2.2. Roll two dice and let $A =$ "The sum is 8" $= \{(2, 6), (3, 5), (4, 4), (5, 3), (6, 2)\}$. $P(A) = 5/36$ since it contains 5 of the 36 possible outcomes.

EXERCISE 2.1. Compute the probability that the sum of the numbers on two dice is k for $2 \leq k \leq 12$.

Example 2.3. Flip three coins and let $B =$ "We get two Heads and one Tail" $= \{(H, H, T), (H, T, H), (T, H, H)\}$. Then $P(B) = 3/8$ since this event contains 3 of the 8 possible outcomes.

We will now derive some basic properties of probabilities and illustrate their use.

(2.1) $$P(A^c) = 1 - P(A)$$

PROOF: Let $A_1 = A$ and $A_2 = A^c$. Then $A_1 \cap A_2 = \emptyset$ and $A_1 \cup A_2 = \Omega$ so (iii) implies $P(A) + P(A^c) = P(\Omega) = 1$ by (ii). Subtracting $P(A)$ from each side of the equation gives (2.1). □

Example 2.4. Flip three coins and let $A = $ "We get at least one Head." It is easy to compute the probability of $A^c = $ "All three coins show Tails" $= \{(T, T, T)\}$. $P(A^c) = 1/8$ since A^c consists of 1 of the 8 possible outcomes. (2.1) then implies $P(A) = 1 - P(A^c) = 7/8$.

$$(2.2) \qquad\qquad P(\emptyset) = 0$$

PROOF: Taking $A = \Omega$ in (2.1), we get $P(\emptyset) = 1 - P(\Omega) = 1 - 1 = 0$ by (ii). □

EXERCISE 2.2. Prove (2.2) by applying (iii) with $A_1 = A_2 = \emptyset$.

(2.3) If $A \subset B$ then $P(A) \leq P(B)$.

PROOF: Let $A_1 = A$ and $A_2 = B - A$. Then $A_1 \cap A_2 = \emptyset$ and $A_1 \cup A_2 = B$, so (iii) implies $P(B) = P(A) + P(B - A) \geq P(A)$ since $P(B - A) \geq 0$. □

(2.4) For any sets A and B,

$$P(A \cup B) = P(A) + P(B) - P(A \cap B)$$

Intuitively, $P(A) + P(B)$ counts $A \cap B$ twice so we have to subtract $P(A \cap B)$ to make the net number of times $A \cap B$ is counted equal to 1.

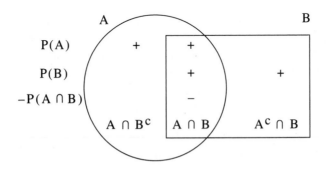

PROOF: To prove this result we just translate the last picture into formulas. We begin by observing that

$$A = (A \cap B) \cup (A \cap B^c) \qquad B = (A \cap B) \cup (A^c \cap B)$$

and in each case the two events on the right are disjoint, so

$$P(A) = P(A \cap B) + P(A \cap B^c)$$
$$P(B) = P(A \cap B) + P(A^c \cap B)$$
$$-P(A \cap B) = -P(A \cap B)$$

Adding the three equalities, we have

$$P(A) + P(B) - P(A \cap B) = P(A \cap B) + P(A \cap B^c) + P(A^c \cap B)$$
$$= P(A \cup B)$$

since $A \cup B = (A \cap B) \cup (A \cap B^c) \cup (A^c \cap B)$ and the three sets on the right-hand side are disjoint. □

Example 2.5. Roll two dice and suppose for simplicity that they are red and green. Let F = "At least one 4 appears," A = "A 4 appears on the red die," and B = "a 4 appears on the green die," so $F = A \cup B$. Now $P(A) = P(B) = 1/6$, since a 4 is 1 of the 6 possibilities when we roll one die, and $A \cap B = \{(4,4)\}$ has probability 1/36, since (4,4) is 1 of the 36 possibilities when we roll two dice. Using (2.4) now,

$$P(A \cup B) = P(A) + P(B) - P(A \cap B) = \frac{1}{6} + \frac{1}{6} - \frac{1}{36} = \frac{11}{36}$$

EXERCISE 2.3. Compute $P(F)$ by computing $P(F^c)$ and using (2.1).

From (2.4) it follows that $P(A \cup B) \leq P(A) + P(B)$. The next result is a generalization to n sets. The intuition behind this inequality is that the right-hand side counts every outcome in the union at least once.

$$(2.5) \qquad P\left(\cup_{i=1}^n A_i\right) \leq \sum_{i=1}^n P(A_i)$$

PROOF: Breaking things down according to the first set an outcome lands in we have

$$\cup_{i=1}^n A_i = A_1 \cup (A_1^c \cap A_2) \cup (A_1^c \cap A_2^c \cap A_3) \ldots \cup (A_1^c \cap \ldots \cap A_{n-1}^c \cap A_n)$$

and the sets on the right-hand side are disjoint so (iii) implies

$$P(\cup_{i=1}^n A_i) = P(A_1) + P(A_1^c \cap A_2) + \ldots + P(A_1^c \cap \ldots \cap A_{n-1}^c \cap A_n)$$
$$\leq P(A_1) + P(A_2) + \ldots + P(A_n)$$

by (2.3) since $A_1^c \cap \ldots \cap A_{j-1}^c \cap A_j \subset A_j$. □

Our final two properties are much less important, but we will need them in the proofs of (2.5)–(2.8) in Chapter 3.

(2.6) Suppose $A_1 \subset A_2 \subset A_3 \ldots$ and $A = \cup_m A_m$. Then

$$P(A_n) \to P(A) \quad \text{as } n \to \infty$$

PROOF: Let $B_n = A_n - A_{n-1}$ for $n = 1, 2, \ldots$ with $A_0 = \emptyset$. The B_n are disjoint with $\cup_{m=1}^n B_m = A_n$ so

$$P(A_n) = \sum_{m=1}^n P(B_m)$$

Letting $n \to \infty$ and recalling that $\sum_{m=1}^\infty c_m$ is defined to be the limit of $\sum_{m=1}^n c_m$, we have

$$\lim_{n \to \infty} P(A_n) = \sum_{m=1}^\infty P(B_m) = P(A)$$

Since the B_m are disjoint with $\cup_{m=1}^\infty B_m = A$. □

(2.7) Suppose $A_1 \supset A_2 \supset A_3 \ldots$ and $A = \cap_m A_m$. Then

$$P(A_n) \to P(A) \quad \text{as } n \to \infty$$

PROOF: Let $E_n = A_n^c$. Then $E_1 \subset E_2 \subset E_3 \ldots$ and Exercise 1.7 implies $A^c = \cup_i E_i$. Using (2.1), (2.6), and then (2.1) again, we have

$$1 - P(A) = P(A^c) = \lim_{n \to \infty} P(E_n)$$
$$= \lim_{n \to \infty} 1 - P(A_n) = 1 - \lim_{n \to \infty} P(A_n)$$

which gives the desired result. □

EXERCISES

2.4. A man receives presents from his three children, Allison, Betty, and Chelsea. To avoid disputes he opens the presents in a random order. What are the possible outcomes?

2.5. Suppose we pick a letter at random from the word TENNESSEE. What is the sample space Ω and what probabilities should be assigned to the outcomes?

2.6. Suppose we pick a number at random from the phone book and look at the last digit. (a) What is the set of outcomes and what probability should be assigned to each outcome? (b) Would this model be appropriate if we were looking at the first digit?

2.7. Suppose we roll a red die and a green die. What is the probability the number on the red die is larger ($>$) than the number on the green die?

2.8. Two dice are rolled. What is the probability (a) the two numbers will differ by 1 or less, (b) the maximum of the two numbers will be 5 or larger?

2.9. If we flip a coin 5 times, what is the probability that the number of heads is an even number (i.e., divisible by 2)?

2.10. Two boys are repeatedly playing a game that they each have probability 1/2 of winning. The first person to win five games wins the match. What is the probability that Al will win if (a) he has won 4 games and Bobby has won 3; (b) he leads by a score of 3 games to 2?

2.11. In Galileo's time people thought that when three dice were rolled, a sum of 9 and a sum of 10 had the same probability since each could be obtained in 6 ways:

$$9 : 1+2+6, \, 1+3+5, \, 1+4+4, \, 2+2+5, \, 2+3+4, \, 3+3+3$$
$$10 : 1+3+6, \, 1+4+5, \, 2+4+4, \, 2+3+5, \, 2+4+4, \, 3+3+4$$

Compute the probabilities of these sums and show that 10 is a more likely total than 9.

2.12. Suppose we roll three dice. Compute the probability that the sum is (a) 3, (b) 4, (c) 5, (d) 6, (e) 7, (f) 8.

2.13. In a group of students, 25% smoke cigarettes, 60% drink alcohol, and 15% do both. What fraction of students have at least one of these bad habits?

2.14. In a group of 320 high school graduates, only 160 went to college but 100 of the 170 men did. How many women did not go to college?

2.15. Suppose $\Omega = \{a, b, c\}$, $P(\{a, b\}) = 0.7$, and $P(\{b, c\}) = 0.6$. Compute the probabilities of $\{a\}$, $\{b\}$, and $\{c\}$.

2.16. Suppose A and B are disjoint with $P(A) = 0.3$ and $P(B) = 0.5$. What is $P(A^c \cap B^c)$?

2.17. Given two events A and B with $P(A) = 0.4$ and $P(B) = 0.7$. What are the maximum and minimum possible values for $P(A \cap B)$?

2.18. Show that $P(A \cap B) \geq P(A) + P(B) - 1$.

2.19. Show that $P(\cap_{i=1}^{n} A_i) \geq 1 - \sum_{i=1}^{n} P(A_i^c)$.

2.20. Show that if $A \subset B$ then $P(B - A) = P(B) - P(A)$.

2.21. Consider the **symmetric difference** $A \Delta B = (A - B) \cup (B - A)$, that is, the outcomes for which exactly one of the two events occurs. Show that $P(A \Delta B) = P(A) + P(B) - 2P(A \cap B)$.

1.3. Permutations and Combinations

In many situations $P(A)$ is the fraction of outcomes that lie in A. To compute probabilities in these situations we have to be able to count the number of outcomes. In this section we will introduce three of the most important counting methods – the multiplication rule, permutations, and combinations – by considering a sequence of examples. The first has already been used in Section 1.1. Before confusing the reader with its precise statement, we give a simple example.

Example 3.1. A man has 4 pair of pants, 6 shirts, 8 pairs of socks, and 3 pairs of shoes. Ignoring the fact that some of the combinations may look ridiculous, he can get dressed in $4 \cdot 6 \cdot 8 \cdot 3 = 576$ ways.

To explain why this is true we begin by considering the number of ways he can put on his pants and shirt. The outcomes of this "experiment" can be written as (i, j) to indicate that the ith pair of pants and the jth shirt were chosen. So the set of outcomes is a table with 4 rows and 6 columns:

$$
\begin{array}{cccccc}
(1,1) & (1,2) & (1,3) & (1,4) & (1,5) & (1,6) \\
(2,1) & (2,2) & (2,3) & (2,4) & (2,5) & (2,6) \\
(3,1) & (3,2) & (3,3) & (3,4) & (3,5) & (3,6) \\
(4,1) & (4,2) & (4,3) & (4,4) & (4,5) & (4,6)
\end{array}
$$

and there are $4 \cdot 6 = 24$ possible outcomes. To prepare for the next level of reasoning, we would like to observe that the same argument shows that if he has n_1 pairs of pants and n_2 shirts then the total number of possibilities is $n_1 \cdot n_2$.

Adding the socks to the problem, we can imagine the possible outcomes as a three-dimensional table, to see that there are $4 \cdot 6 \cdot 8$ possibilities, or, building on the last answer, we can note that there are 24 ways of putting on pants and shirt and 8 ways of putting on socks, so there are $192 = 24 \cdot 8 = 4 \cdot 6 \cdot 8$ ways of putting on pants, shirt, and socks. Repeating the last argument one more time, we see that there are 192 ways to put on pants, shirt, and socks, and 3 ways to put on shoes, so there are $192 \cdot 3 = 4 \cdot 6 \cdot 8 \cdot 3$ ways that the man in our problem can get dressed.

The reasoning in the last solution can clearly be extended to more than four experiments, and does not depend on the number of choices at each stage, so we have

(3.1) **The multiplication rule.** Suppose that m experiments are performed in order and that, no matter what the outcomes of experiments $1, \ldots, k-1$ are, experiment k has n_k possible outcomes. Then the total number of outcomes is $n_1 \cdot n_2 \cdots n_m$.

Since it is important for our first application, we would like to point out that the actual set of choices at the kth stage may depend upon what happened on earlier choices, as long as the number of choices does not. For practice applying the multiplication rule, do

EXERCISE 3.1. A restaurant offers soup or salad to start, and has 11 entrees to choose from, each of which is served with rice, baked potato, or zucchini. How many meals can you have if you can choose to eat one of their 4 desserts or have no dessert?

EXERCISE 3.2. How many answer sheets are possible for a true/false test with 15 questions?

To introduce our next counting result we consider

Example 3.2. How many ways can 5 people stand in line?

To answer this question, we think about building the line up one person at a time starting from the front. There are 5 people we can choose to put at the front of the line. Having made the first choice, we have 4 possible choices for the second position. (The set of people we have to choose from depends upon who was chosen first, but there are always 4 people to choose from.) Continuing, there are 3 choices for the third position, 2 for the fourth, and finally 1 for the last. Invoking the multiplication rule, we see that the answer must be

$$5 \cdot 4 \cdot 3 \cdot 2 \cdot 1 = 120$$

Generalizing from the last example we define n **factorial** to be

$$n! = n \cdot (n-1) \cdot (n-2) \cdots 2 \cdot 1$$

To see that this gives the number of ways n people can stand in line, notice that there are n choices for the first person, $n-1$ for the second, and each subsequent choice reduces the number of people by 1 until finally there is only

1 person who can be the last in line. For practice with applying this reasoning, try

EXERCISE 3.3. How many different batting orders are possible for 9 baseball players?

EXERCISE 3.4. How many ways can 8 books be put on a shelf?

To evaluate your answers to the last two problems, use the following table. Note that $n!$ grows very quickly since $n! = n \cdot (n-1)!$.

1!	1	5!	120	9!	362,880
2!	2	6!	720	10!	3,628,800
3!	6	7!	5,040	11!	39,916,800
4!	24	8!	40,320	12!	479,001,600

To connect with the previous topic, our next question can be thought of as: "How many ways can we make a line of k people out of a group of $n > k$?" The next situation is an example of this with $k = 4$ and $n = 12$, but the story is slightly different.

Example 3.3. Twelve people belong to a club. How many ways can they pick a president, vice-president, secretary, and treasurer?

Again we think of filling the offices one at a time in the order in which they were given in the last sentence. There are 12 people we can pick for president. Having made the first choice, there are always 11 possibilities for vice-president, 10 for secretary, and 9 for treasurer. So by the multiplication rule, the answer is

$$\frac{12\ 11\ 10\ 9}{P\ V\ S\ T}$$

Passing to the general situation, if we have k offices and n club members then the answer is

$$(3.2) \qquad n \cdot (n-1) \cdot (n-2) \cdots (n-k+1)$$

To see this, note that there are n choices for the first office, $n-1$ for the second, and so on until there are $n-k+1$ choices for the last, since after the last person is chosen there will be $n-k$ left. Products like the last one come up so often that they have a name: the **permutation of n things taken k at a time,** or $P_{n,k}$ for short. Multiplying and dividing by $(n-k)!$ we have

$$n \cdot (n-1) \cdot (n-2) \cdots (n-k+1) \cdot \frac{(n-k)!}{(n-k)!} = \frac{n!}{(n-k)!}$$

which gives us a short formula,

(3.3) $$P_{n,k} = n!/(n-k)!$$

Formula (3.3) will be useful in a minute. In solving problems, it is easier to use (3.2). For practice with permutations, try

EXERCISE 3.5. In a horse race the first three finishers are said to win, place, and show. How many finishes are possible for a race with 11 horses?

EXERCISE 3.6. Five different awards are to be given to a class of 30 students. How many ways can this be done if (a) each student can receive any number of awards, (b) each student can receive at most one award?

In the problems just considered, the order in which the choices were made was important. In the next set of problems the order in which the choices are made is not important. We only care about the set of things chosen.

Example 3.4. A club has 23 members. How many ways can they pick 4 people to be on a committee to plan a party?

To reduce this question to the previous situation, we imagine making the committee members stand in line, which by (3.2) can be done in $23 \cdot 22 \cdot 21 \cdot 20$ ways. To get from this to the number of committees, we note that each committee can stand in line 4! ways, so the number of committees is the number of lineups divided by 4! or

$$\frac{23 \cdot 22 \cdot 21 \cdot 20}{1 \cdot 2 \cdot 3 \cdot 4} = 23 \cdot 11 \cdot 7 \cdot 5 = 8,855$$

Passing to the general situation, suppose we want to pick k people out of a group of n. Our first step is to make the k people stand in line, which can be done in $P_{n,k}$ ways, and then to realize that each set of k people can stand in line $k!$ ways, so the number of ways to choose k people out of n is

(3.4) $$C_{n,k} = \frac{P_{n,k}}{k!} = \frac{n!}{k!(n-k)!} = \frac{n \cdot (n-1) \cdots (n-k+1)}{1 \cdot 2 \cdots k}$$

by (3.3) and (3.2). Here, $C_{n,k}$ is short for the number of **combinations of n things taken k at a time.** $C_{n,k}$ is often written as $\binom{n}{k}$, a symbol that is read as "n choose k." We will use both of these notations throughout the book. For practice with combinations, try

EXERCISE 3.7. A restaurant offers 15 possible toppings for its pizzas. How many different pizzas with 4 different toppings can be ordered?

EXERCISE 3.8. We are going to pick 5 cards out of a deck of 52. In how many ways can this be done?

Our next result shows that the $C_{n,k}$ are also important outside of probability theory.

The Binomial Theorem. If we want to evaluate the product

$$(a+b)^4 = (a+b)(a+b)(a+b)(a+b)$$

then there are $2^4 = 16$ terms, since we can pick an a or b from within each pair of parentheses. To compute $(a+b)^4$ we begin by noting that there is only 1 way to pick all a's, so the coefficient of a^4 will be 1. Turning to a^3b, there are $C_{4,3} = 4!/(3!1!) = 4$ ways of picking 3 a's (and hence 1 b) from the four pairs of parentheses, so the second term is $4a^3b$. In a similar way we can see that there are $C_{4,2} = 4!/(2!2!) = (4 \cdot 3)/2 = 6$ ways of picking 2 a's and 2 b's, $C_{4,1} = 4$ ways of picking 1 a and 3 b's, and only one way to pick 4 b's. Thus

$$(a+b)^4 = a^4 + 4a^3b + 6a^2b^2 + 4ab^3 + b^4$$

Generalizing from the last example, we see that

(3.5)
$$(a+b)^n = \sum_{k=0}^{n} \binom{n}{k} a^k b^{n-k}$$

since there are $C_{n,k}$ ways of picking k a's and $n-k$ b's from the n terms in the product. To make the last equation valid we define

(3.6)
$$0! = 1 \quad \text{so that} \quad \binom{n}{0} = \binom{n}{n} = \frac{n!}{n!0!} = 1$$

Because of (3.5), the numbers $C_{n,k} = \binom{n}{k}$ are sometimes called **binomial coefficients**.

EXERCISE 3.9. Show that (a) $\sum_{m=0}^{n} \binom{n}{m} = 2^n$; (b) $\sum_{m=0}^{n} (-1)^m \binom{n}{m} = 0$.

EXERCISE 3.10. Find (a) $(x+2)^5$, (b) $(2x+3)^3$.

The binomial coefficients fit together in a pretty pattern called **Pascal's triangle**. To derive the relationship that generates Pascal's triangle, note that when we are picking k things out of n, which can be done in $C_{n,k}$ ways, then we may or may not pick the last object. If we pick the last object then we must complete our set of k by picking $k-1$ objects from the first $n-1$, which can be done in $C_{n-1,k-1}$ ways. If we do not pick the last object then we must pick all k objects from the first $n-1$, which can be done in $C_{n-1,k}$ ways. Combining the last three sentences, we have

$$(3.7) \qquad\qquad C_{n,k} = C_{n-1,k-1} + C_{n-1,k}$$

To make the last result hold for $k=0$ and $k=n$ we introduce the convention that

$$(3.8) \qquad\qquad C_{n,m} = 0 \quad \text{if } m < 0 \text{ or } m > n$$

If we write down $C_{n,k}$ with $0 \le k \le n$ on the nth row in just the right way we get a triangle in which each value is the sum of the two closest numbers on the row above it.

$$
\begin{array}{ccccccccccccccc}
& & & & & & & 1 & & & & & & & \\
& & & & & & 1 & & 1 & & & & & & \\
& & & & & 1 & & 2 & & 1 & & & & & \\
& & & & 1 & & 3 & & 3 & & 1 & & & & \\
& & & 1 & & 4 & & 6 & & 4 & & 1 & & & \\
& & 1 & & 5 & & 10 & & 10 & & 5 & & 1 & & \\
& 1 & & 6 & & 15 & & 20 & & 15 & & 6 & & 1 & \\
1 & & 7 & & 21 & & 35 & & 35 & & 21 & & 7 & & 1 \\
\end{array}
$$

Note that the table is symmetric, i.e.,

$$(3.9) \qquad\qquad C_{n,m} = C_{n,n-m}$$

The last equality is easy to prove: The number of ways of picking m objects out of n to take is the same as the number of ways of choosing $n-m$ to leave behind. Of course, one can also see (3.9) directly from the formula in (3.4):

$$C_{n,n-m} = \frac{n!}{(n-m)!(n-(n-m))!} = \frac{n!}{(n-m)!m!} = C_{n,m}$$

EXERCISE 3.11. Prove that $\binom{n}{r+1} = \frac{n-r}{r+1}\binom{n}{r}$ and use this relationship to compute the binomial coefficients with $n=10$.

We defined $C_{n,k}$ as the number of ways of picking k things out of n. To motivate the next generalization we would like to observe that $C_{n,k}$ is also the number of ways we can divide n objects into two piles, one pile with k objects and the other with $n - k$. To connect this observation with the next problem, think of it as asking: "How many ways can we divide 12 objects into three piles of sizes 4, 3, and 5?"

Example 3.5. A house has 12 rooms. We want to paint 4 yellow, 3 purple, and 5 red. In how many ways can this be done?

To generate all the possibilities, we can first decide the order in which the rooms will be painted, which can be done in 12! ways, then paint the first 4 on the list yellow, the next 3 purple, and the last 5 red. One example is

$$\frac{9}{Y}\ \frac{6}{Y}\ \frac{11}{Y}\ \frac{1}{Y}\ \frac{8}{P}\ \frac{2}{P}\ \frac{10}{P}\ \frac{5}{R}\ \frac{3}{R}\ \frac{7}{R}\ \frac{12}{R}\ \frac{4}{R}$$

Now, the first four choices can be rearranged in 4! ways without affecting the outcome, the middle three in 3! ways, and the last five in 5! ways. Invoking the multiplication rule, we see that in a list of the 12! possible permutations each possible painting thus appears 4! 3! 5! times. Hence the number of possible paintings is

$$\frac{12!}{4!\,3!\,5!} = \frac{12 \cdot 11 \cdot 10 \cdot 9 \cdot 8 \cdot 7 \cdot 6}{1 \cdot 2 \cdot 3 \cdot 1 \cdot 2 \cdot 3 \cdot 4} = 27,720$$

Another way of getting the last answer is to first pick 4 of the 12 rooms to be painted yellow, which can be done in $C_{12,4}$ ways, and then pick 3 of the remaining 8 rooms to be painted purple, which can be done in $C_{8,3}$ ways. (The 5 unchosen rooms will be painted red.) This is the same answer since

$$C_{12,4}C_{8,3} = \frac{12!}{4!\,8!} \cdot \frac{8!}{3!\,5!} = \frac{12!}{4!\,3!\,5!}.$$

The second computation is simpler but the first one makes it easier to see

(3.10) If we have a group of n objects to be divided into m groups of size $n_1, \ldots, n_m$ with $n_1 + \cdots + n_m = n$ this can be done in

$$\frac{n!}{n_1!\,n_2! \cdots n_m!}\ \text{ways.}$$

For practice applying this formula try

EXERCISE 3.12. There are 37 students in a class. In how many ways can a professor give out 3 A's, 4 B's, 5 C's, and 25 F's?

EXERCISE 3.13. A child has 15 blocks: 6 red, 4 yellow, and 5 blue. How many ways can they be put in a line?

EXERCISE 3.14. Four people play a card game in which each gets 13 cards. How many possible deals are there?

The numbers in (3.10) are called **multinomial coefficients** since an extension of the argument for (3.5) shows that

$$(3.11) \qquad (x_1 + \cdots + x_m)^n = \sum \frac{n!}{n_1! \, n_2! \cdots n_m!} x_1^{n_1} \cdots x_m^{n_m}$$

where the sum is over all integers $n_1, \ldots, n_m \geq 0$ with $n_1 + \cdots + n_m = n$. Even simple examples of (3.11) are lengthy to write out:

$$(x + y + z)^3 = x^3 + y^3 + z^3$$
$$+ \, 3x^2 y + 3x^2 z + 3y^2 x + 3y^2 z + 3z^2 x + 3z^2 y + 6xyz$$

EXERCISES

3.15. Five businessmen meet at a convention. How many handshakes are exchanged if each shakes hands with all the others?

3.16. (a) How many license plates are possible if the first three places are occupied by letters and the last three by numbers? (b) Assuming all combinations are equally likely, what is the probability the three letters and the three numbers are different?

3.17. A domino is an ordered pair (m, n) with $0 \leq m \leq n \leq 6$. How many dominoes are in a set if there is only one of each?

3.18. A basketball team has 5 players over six feet tall and 6 who are under six feet. How many ways can they have their picture taken if the 5 taller players stand in a row behind the 6 shorter players who are sitting on a row of chairs?

3.19. Six students, three boys and three girls, lineup in a random order for a photograph. What is the probability that the boys and girls alternate?

3.20. Seven people sit at a round table. How many ways can this be done if Mr. Jones and Miss Smith (a) must sit next to each other, (b) must not sit next to each other?

3.21. How many ways can 4 couples sit at a round table if each man sits next to two women, neither one of whom is his wife?

3.22. Suppose that n people are seated at random in a row of n seats. What is the probability Mr. Jones and Miss Smith sit next to each other?

3.23. Let $B_{n,k}$ be the number of ways that k people can sit in a row of n chairs with no two sitting next to each other. (a) Show that

$$B_{n,k+1} = \sum_{m=2k-1}^{n-2} B_{m,k}$$

(b) Use the last relationship to compute $B_{n,2}$ and $B_{8,3}$.

3.24. How many ways can 4 men and 4 women sit in a row if no two men or two women sit next to each other?

3.25. How many different ways can the letters in the following words be arranged: (a) money, (b) banana, (c) statistics, (d) mississippi?

3.26. Twelve different toys are to be divided among 3 children so that each one gets 4 toys. How many ways can this be done?

3.27. A club with 50 members is going to form two committees, one with 8 members and the other with 7. How many ways can this be done (a) if the committees must be disjoint? (b) if they can overlap?

3.28. If seven dice are rolled, what is the probability that each of the six numbers will appear at least once?

3.29. How many ways can 5 history books, 3 math books, and 4 novels be arranged on a shelf if the books of each type must be together?

3.30. Suppose three runners from team A and three runners from team B have a race. If all six runners have equal ability, what is the probability that the three runners from team A will finish first, second, and fourth?

3.31. (a) Eight people are divided into four pairs to play bridge. In how many ways can this be done? (b) Suppose there are 4 men and 4 women and we insist that each pair contain a man and a woman.

3.32. By considering the selection of k students from a class with m boys and n girls, show that (recall that $\binom{j}{i} = 0$ if $i > j$)

$$\binom{n+m}{k} = \binom{n}{0}\binom{m}{k} + \binom{n}{1}\binom{m}{k-1} + \cdots \binom{n}{k}\binom{m}{0}$$

Taking $k = m = n$ and using $\binom{n}{n-\ell} = \binom{n}{\ell}$ gives

$$\binom{2n}{n} = \sum_{\ell=0}^{n} \binom{n}{\ell}^2$$

3.33. By considering the position of the leftmost object selected, show that

$$\binom{n}{m} = \binom{n-1}{m-1} + \binom{n-2}{m-1} + \ldots + \binom{m-1}{m-1}$$

This says that any entry in Pascal's triangle is the sum of those on the diagonal line above it.

3.34. Show that $\sum_{k=1}^{n} k\binom{n}{k} = n2^{n-1}$ by considering the number of ways we can pick a committee with $k \geq 1$ members including one special member called the chairman.

3.35. Show that $\sum_{k=1}^{n} k(k-1)\binom{n}{k} = n(n-1)2^{n-1}$.

3.36. Prove the last two identities by applying the binomial theorem to $(1+x)^n$, differentiating with respect to x and setting $x = 1$.

3.37. Use the approach in the last problem to show that $\sum_{k=1}^{n}(-1)^{k-1}k\binom{n}{k} = 0$.

Physical Particles.* If n tourists pick from m hotels at random then (3.10) tells us that the probability that exactly n_1 will end up in hotel 1, n_2 in hotel 2, ..., and n_m in hotel m is

$$\frac{n!}{n_1! n_2! \cdots n_m!} m^{-n}$$

These probabilities, called **Maxwell-Boltzmann statistics**, come up in a number of probability problems. However, in most situations in physics one deals with indistinguishable particles that follow so-called **Bose-Einstein statistics**. To count the number of ways that n indistinguishable particles can be divided into m groups is easy when $m = 2$. An outcome is described completely by the number of balls n_1 in the first box (for then $n_2 = n - n_1$) and we must have $0 \leq n_1 \leq n$, so there are $n + 1$ possibilities. The formula for $m > 2$ groups requires a trick that is motivated by the solution of the following related problem.

Example 3.6. How many ways can we divide n indistinguishable balls into m groups in such a way that there is at least one ball per group?

To solve this problem we imagine a line of n white balls and $m - 1$ pieces of cardboard to place between them to indicate the boundaries of the groups. An example with $n = 14$ and $m = 6$ would be

$$ooo|o|oo|oooo|ooo|o$$

---

* Material that comes after the end-of-section exercises have begun can be omitted without loss.

Since we must pick $m - 1$ of the $n - 1$ spaces to put our cardboard dividers into, there are $C_{n-1,m-1}$ possibilities.

When the groups can have size 0, the last scheme breaks down and we need to look at things in a different way. We imagine having n white balls (o's) and $m - 1$ black balls (x's) that will indicate the boundaries between groups. An example with $n = 9$, $m = 6$ would be

$$o \, o \, x \, x \, o \, o \, o \, o \, x \, o \, x \, o \, o \, x$$

which corresponds to groups of size $n_1 = 2$, $n_2 = 0$, $n_3 = 4$, $n_4 = 1$, $n_5 = 2$, and $n_6 = 0$. Each possible division of n indistinguishable balls into m boxes corresponds to one arrangement of the n white balls and $m - 1$ black balls, which is in turn the same as the number of ways of getting n heads and $m - 1$ tails in $n + m - 1$ tosses, so there are $C_{n+m-1,m-1}$ outcomes.

Example 3.7. How many different partial derivatives of order 3 are there for a function of 3 variables? (Recall that for a partial derivative the only thing that is important is the number of times we differentiate with respect to each variable. The order in which the derivatives are taken is unimportant.)

Here $n = 3$, $m = 3$, so our formula says the answer is $C_{5,2} = (5 \cdot 4)/2 = 10$. Writing (i, j, k) for $\partial^3 f / \partial x_1^i \partial x_2^j \partial x_3^k$, we can check the answer by enumerating the possibilities: $(3, 0, 0)$, $(0, 3, 0)$, $(0, 0, 3)$, $(2, 1, 0)$, $(2, 0, 1)$, $(1, 2, 0)$, $(0, 2, 1)$, $(1, 0, 2)$, $(0, 1, 2)$, $(1, 1, 1)$.

Example 3.8. Suppose we place 14 indistinguishable balls in 7 boxes. What is the probability that there will be at least one ball per box?

As we have already seen, the total number of outcomes is $C_{n+m-1,m-1} = C_{20,6}$, while the number of outcomes with at least one ball per box is $C_{n-1,m-1} = C_{13,6}$, so the probability of interest is

$$\frac{C_{13,6}}{C_{20,6}} = \frac{13 \cdot 12 \cdot 11 \cdot 10 \cdot 9 \cdot 8}{20 \cdot 19 \cdot 18 \cdot 17 \cdot 16 \cdot 15} = \frac{1,235,520}{27,907,200} = 0.04427$$

Notice that the probability is small even though there are twice as many balls as boxes.

3.38. Indistinguishable particles are said to obey **Fermi-Dirac** statistics if all arrangements that have at most one particle per box have the same probability. How many ways can we put m of these particles in $n \geq m$ boxes?

3.39. Suppose we put n indistinguishable balls into m boxes. What is the probability there will be k balls in the first box?

3.40. Suppose we put n indistinguishable balls into m boxes. What is the probability there will be exactly j empty boxes?

3.41. How many ways can 5 people sit in a row of 18 seats if no two people sit next to each other?

3.42. An elevator starts in the basement with 10 people (not including the elevator operator) and each person gets off at one of the six floors. How many outcomes are there if all people look the same?

1.4. Urn Problems

In many situations we pick at random from a finite set of objects of several different types. A convenient mathematical idealization of such experiments is picking from an urn of colored balls.

Example 4.1. Suppose we pick 4 balls out of an urn with 12 red balls and 8 black balls. What is the probability of $B =$ "We get two balls of each color"?

Almost by definition, there are

$$\binom{20}{4} = \frac{20 \cdot 19 \cdot 18 \cdot 17}{1 \cdot 2 \cdot 3 \cdot 4} = 5 \cdot 19 \cdot 3 \cdot 17 = 4,845$$

ways of picking 4 balls out of the 20. To count the number of outcomes in B, we note that there are $C_{12,2}$ ways to choose the red balls and $C_{8,2}$ ways to choose the black balls, so the multiplication rule implies

$$|B| = \binom{12}{2}\binom{8}{2} = \frac{12 \cdot 11}{1 \cdot 2} \cdot \frac{8 \cdot 7}{1 \cdot 2} = 6 \cdot 11 \cdot 4 \cdot 7 = 1,848$$

It follows that $P(B) = 1848/4845 = 0.3814$.

Our next example is a situation in which people do actually draw balls out of an urn.

Example 4.2. New York State Lotto. In this lottery game, you pick six of the numbers 1 through 54, and then in a televised drawing six of these numbers are selected. If all six of your numbers are selected then you win a share of the first prize. (50% of the prize pool = 19% of the money bet, is divided among all people who pick the six numbers correctly.) If five or four of your numbers are selected you win a share of the second or third prize (11% and 28% of the prize pool respectively). What is the probability you will win a share of the first prize? second prize? third prize?

There are
$$C_{54,6} = \frac{54 \cdot 53 \cdot 52 \cdot 51 \cdot 50 \cdot 49}{1 \cdot 2 \cdot 3 \cdot 4 \cdot 5 \cdot 6} = 25,827,165$$

ways of picking 6 numbers, so the probability of winning a share of the first prize is $1/25,827,165$. To calculate the probability of winning a share of the second prize we observe that there are $C_{6,5} \cdot C_{48,1} = 6 \cdot 48 = 288$ ways they can select 6 numbers that include 5 of your 6 and 1 of the remaining 48, so your probability of winning a share of the second prize is

$$\frac{288}{25,827,165} \approx \frac{1}{89,678}$$

Third prize is much easier to win since there are $C_{6,4}C_{48,2} = 15 \cdot 1128 = 16,920$ ways they can pick 4 of your 6 numbers and 2 of the remaining 48, giving a probability of
$$\frac{16,920}{25,827,165} \approx \frac{1}{1526}$$

for winning a share of the third prize.

EXERCISE 4.1. In the New York State Lotto, you actually pick six numbers and a seventh "supplemental" number from 1 through 54, and then six numbers are selected in a televised drawing. If your supplemental number and three of your six numbers are selected, you win a share of the fourth prize (the remaining 11% of the prize pool). What is the probability you win a share of the fourth prize?

Our next three examples are practical applications of drawing balls out of urns.

Example 4.3. Disputed elections. In a close election in a small town, 2,656 people voted for candidate A compared to 2,594 who voted for candidate B, a margin of victory of 62 votes. An investigation of the election, instigated no doubt by the loser, found that 136 of the people who voted in the election should not have. Since this is more than the margin of victory, should the election results be thrown out even though there was no evidence of fraud on the part of the winner's supporters?

Like many problems that come from the real world (*DeMartini v. Power*, 262 NE2d 857), this one is not precisely formulated. To turn this into a probability problem we suppose that all the votes were equally likely to be one of the 136 erroneously cast and we investigate what happens when we remove 136 balls from an urn with 2,656 white balls and 2,594 black balls. Now the probability of removing exactly m white and $136 - m$ black balls is

$$\frac{\binom{2,656}{m}\binom{2,594}{136-m}}{\binom{5,250}{136}}$$

In order to reverse the outcome of the election, we must have

$$2,656 - m \leq 2,594 - (136 - m) \quad \text{or} \quad m \geq 99$$

With the help of a short computer program we can sum the probability above from $m = 99$ to 136 to conclude that the probability of the removal of 136 randomly chosen votes reversing the election is 7.492×10^{-8}. This computation supports the Court of Appeals decision to overturn a lower court ruling that voided the election in this case.

EXERCISE 4.2. What do you think should have been done in *Ipolito v. Power*, 241 NE2d 232, where the winning margin was 1,422 to 1,405 but 101 votes had to be thrown out?

Example 4.4. Acceptance sampling of a manufactured product. Suppose we have a lot of M light bulbs, which contains m defective bulbs. If we pick N for testing, what is the probability that no bulbs in the sample are defective?

The answer is simple

$$\frac{C_{M-m,N}}{C_{M,N}} = \frac{P_{M-m,N}/N!}{P_{M,N}/N!} = \frac{(M-m) \cdot (M-m-1) \cdots (M-m-N+1)}{M \cdot (M-1) \cdots (M-N+1)}$$

The denominator on the left-hand side gives the number of ways we can pick our sample of size N assuming the order of choice is not important, and the numerator gives the number of ways we can pick N from the $M-m$ nondefective items. As the right-hand side indicates, we would have gotten the same answer, with slightly less arithmetic, if we had used a sample space in which the order of choice was important.

Suppose, for a concrete example, that $M = 1000$ and we pick a sample of size $N = 60$. If there were $m = 50$ defective bulbs then the probability of no defective bulbs in our sample would be

$$\frac{950 \cdot 949 \cdots 891}{1000 \cdot 999 \cdots 941} \approx (0.95)^{60} = 0.04607$$

since $0.95 = (950/1000) > (949/999) \cdots > (891/941) = 0.9468$. Thus if we sample 60 bulbs and find no defectives the probability of 50 or more defective bulbs in the lot of 1000 is about 0.0464.

Generalizing from the last computation, suppose that N is much smaller than M, which is the usual situation, and we let $p = m/M$. If we test N units and find no defectives then the probability a fraction p is defective is about $(1-p)^N$. From this we see that the number of items we should choose to test depends upon the level of defectives p that we want to be able to detect and does not depend upon the size of the population M.

Example 4.5. Capture-recapture experiments. An ecology graduate student goes to a pond and captures $k = 60$ water beetles, marks each with a dot of paint, and then releases them. A few days later she goes back and captures another sample of $r = 50$, finding $m = 12$ marked beetles and $r - m = 38$ unmarked. What is her best guess about the size of the population of water beetles?

To turn this into a precisely formulated problem, we will suppose that no beetles enter or leave the population between the two visits. With this assumption, if there were N water beetles in the pond, then the probability of getting m marked and $r - m$ unmarked in a sample of r would be

$$p_N = \frac{C_{k,m} \, C_{N-k,r-m}}{C_{N,r}}$$

To estimate the population we will pick N to maximize p_N, the so-called **maximum likelihood estimate**. To find the maximizing N, we note that

$$C_{j-1,i} = \frac{(j-1)!}{(j-i-1)!i!} \quad \text{so} \quad C_{j,i} = \frac{j!}{(j-i)!i!} = \frac{jC_{j-1,i}}{(j-i)}$$

and it follows that

$$p_N = p_{N-1} \cdot \frac{N-k}{N-k-(r-m)} \cdot \frac{N-r}{N}$$

Now $p_N/p_{N-1} \geq 1$ if and only if

$$(N-k)(N-r) \geq N(N-k-r+m)$$

that is,

$$N^2 - kN - rN + kr \geq N^2 - kN - rN + mN$$

or equivalently if $kr \geq Nm$ or $N \leq kr/m$. Thus the value of N that maximizes the probability p_N is the largest integer $\leq kr/m$. This choice is reasonable since when $N = kr/m$ the proportion of marked beetles in the population $k/N = m/r$, the proportion of marked beetles in the sample. Plugging in the numbers from our example, $kr/m = (60 \cdot 50)/12 = 250$, so the probability is maximized when $N = 250$.

We close this section by returning to our gambling roots.

Example 4.6. Poker. In the game of poker the following hands are possible; they are listed in increasing order of desirability. In the definitions the word *value* refers to A, K, Q, J, 10, 9, 8, 7, 6, 5, 4, 3, or 2. This sequence also describes the relative ranks of the cards, with one exception: an Ace may be regarded as a 1 for the purposes of making a straight. (See the example in (d), below.)

(a) *one pair:* two cards of equal value plus three cards with different values

$\quad$ J♠ J♢ 9♡ Q♣ 3♠

(b) *two pair:* two pairs plus another card with a different value

$\quad$ J♠ J♢ 9♡ 9♣ 3♠

(c) *three of a kind:* three cards of the same value and two with different values

$\quad$ J♠ J♢ J♡ 9♣ 3♠

(d) *straight:* five cards with consecutive values

$\quad$ 5♡ 4♠ 3♠ 2♡ A♣

(e) *flush:* five cards of the same suit

$\quad$ K♣ 9♣ 7♣ 6♣ 3♣

(f) *full house:* a three of a kind and a pair

$\quad$ J♠ J♢ J♡ 9♣ 9♠

(g) *four of a kind:* four cards of the same value plus another card

$\quad$ J♠ J♢ J♡ J♣ 9♠

(h) *straight flush:* five cards of the same suit with consecutive values

$\quad$ A♣ K♣ Q♣ J♣ 10♣

This example is called a *royal flush.*

To compute the probabilities of these poker hands we begin by observing that there are

$$\binom{52}{5} = \frac{52 \cdot 51 \cdot 50 \cdot 49 \cdot 48}{1 \cdot 2 \cdot 3 \cdot 4 \cdot 5} = 2,598,960$$

ways of picking 5 cards out of a deck of 52, so it suffices to compute the number of ways each hand can occur. We will do three cases to illustrate the main ideas and then leave the rest to the reader.

(d) *straight:* $10 \cdot 4^5$

A straight must start with a card that is 5 or higher, 10 possibilities. Once the values are decided on, suits can be assigned in 4^5 ways. This counting regards a straight flush as a straight. If you want to exclude straight flushes, suits can be assigned in $4^5 - 4$ ways.

(f) *full house:* $13 \cdot C_{4,3} \cdot 12 \cdot C_{4,2}$

We first pick the value for the three of a kind (which can be done in 13 ways), then assign suits to those three cards ($C_{4,3}$ ways), then pick the value for the pair (12 ways), then we assign suits to the last two cards ($C_{4,2}$ ways).

(a) *one pair:* $13 \cdot C_{4,2} \cdot C_{12,3} \cdot 4^3$

We first pick the value for the pair (13 ways), next pick the suits for the pair ($C_{4,2}$ ways), then pick three values for the other cards ($C_{12,3}$ ways) and assign suits to those cards (in 4^3 ways).

A common incorrect answer to this question is $13 \cdot C_{4,2} \cdot 48 \cdot 44 \cdot 40$. The faulty reasoning underlying this answer is that the third card must not have the same value as the cards in the pair (48 choices), the fourth must be different from the third and the pair (44 choices), ... However, this reasoning is flawed since it counts each outcome $3! = 6$ times. (Note that $48 \cdot 44 \cdot 40/3! = C_{12,3} \cdot 4^3$.)

EXERCISE 4.3. Compute the probabilities of (b) two pair, (c) three of a kind, (e) a flush, (g) four of a kind, (h) a straight flush.

The numerical values of the probabilities of all poker hands are given in the next table.

(a) *one pair*	.422569
(b) *two pair*	.047539
(c) *three of a kind*	.021128
(d) *straight*	.003940
(e) *flush*	.001981
(f) *full house*	.001441
(g) *four of a kind*	.000240
(h) *straight flush*	.000015

The probability of getting none of these hands can be computed by summing the values for (a) through (g) (recall that (d) includes (h)) and subtracting the result from 1. However, it is much simpler to observe that we have nothing if we have five different values that do not make a straight or a flush. So the number of nothing hands is $(C_{13,5} - 10) \cdot (4^5 - 4)$ and the probability of a nothing hand is 0.501177.

EXERCISES

4.4. Two red cards and two black cards are lying face down on the table. You pick two cards and turn them over. What is the probability that the two cards are different colors?

4.5. Four people are chosen at random from 5 couples. What is the probability two men and two women are selected?

4.6. You pick 5 cards out of a deck of 52. What is the probability you get exactly 2 spades?

4.7. Seven students are chosen at random from a class with 17 boys and 13 girls. What is the probability that 4 boys and 3 girls are selected?

4.8. In a carton of 12 eggs, 2 are rotten. If we pick 4 eggs to make an omelet, what is the probability we do not get a rotten egg?

4.9. Pick 6 cards out of a deck of 52. What is the probability that there is at least one pair?

4.10. A closet contains 8 pairs of shoes. You pick out 5. What is the probability of (a) no pair, (b) exactly one pair, (c) two pairs?

4.11. A drawer contains 10 black, 8 brown, and 6 blue socks. If we pick two socks at random, what is the probability they match?

4.12. Two cards are a blackjack if one is an A and the other is a K, Q, J, or 10. If you pick two cards out of a deck, what is the probability you will get a blackjack?

4.13. In Keno the casino picks 20 balls from a set of 80 numbered 1 through 80. Before this drawing is done, you pick 10 numbers. What is the probability that exactly 5 of your numbers will be in the 20 selected?

4.14. A student studies 12 problems from which the professor will randomly choose 6 for a test. If the student can solve 9 of the problems, what is the probability she can solve at least 5 of the problems on the test?

4.15. An urn contains 6 red balls and 6 black balls. We draw without replacement from this urn until a red ball is chosen. Let K be the number of balls chosen. Compute $P(K = k)$ for $k = 1, 2, 3, 4, 5, 6, 7$. The number of computations required can be reduced considerably by relating $P(K = k)$ to $P(K = k - 1)$.

4.16. A football team has 16 seniors, 12 juniors, 8 sophomores, and 4 freshmen. If we pick 5 players at random, what is the probability we will get 2 seniors and 1 from each of the other 3 classes?

4.17. Suppose we pick 5 cards out of a deck of 52. What is the probability we get at least one card of each suit?

1.5. Repeated Experiments

The common feature of all the problems in this section is that we repeatedly choose at random from a finite set of possibilities, but the choices made earlier do not reduce the number of choices available at the kth stage. In terms of urns, we are "drawing with replacement." That is, we pick a ball, note its color and return it to the urn. This is in contrast to the urn problems in the last section, which involve "drawing without replacement."

Example 5.1. The Birthday Problem. There are 30 students in a probability class. What is the probability of $A = $ "No two students have the same birthday"?

To answer this question we will suppose that leap year does not exist and that all birthdays are equally likely, so that if we list the students in alphabetical order all $(365)^{30}$ possible sequences of birthdays are equally likely. To evaluate $P(A)$ we have to count the number of outcomes in A. To do this, we think of building up the outcomes in A one birthday at a time. The first student can have any of the 365 birthdays. The second student must avoid the first student's birthday and hence has 364 choices. Continuing in this way, each student has one fewer choice until the 30th student must avoid the previous 29 birthdays and has 336 choices. Applying the multiplication rule, we see that

$$|A| = 365 \cdot 364 \cdots 336$$

The last answer can also be obtained by noting that there are 365 birthdays and we are going to pick 30 of them to "stand in line," so the number of possibilities is $P_{365,30} = 365!/335!$. To compute $P(A)$ now we observe that

$$P(A) = \frac{|A|}{|\Omega|} = \frac{365}{365} \cdot \frac{364}{365} \cdots \frac{336}{365} = 0.2937$$

At first sight the last answer may be surprising: 30 is less than 1/12 of the 365 days on the calendar, but more than 70% of the time two people will have the same birthday. This "paradox" evaporates when we realize that we have $C_{30,2} = (30 \cdot 29)/2 = 435$ pairs of people, each of whom has a common birthday with probability 1/365.

Let p_k be the probability that k people all have different birthdays. Clearly, $p_1 = 1$ and $p_{k+1} = p_k(365 - k)/365$. Using this recursion it is easy to generate a table of p_k for $1 \le k \le 40$:

1	1.00000	11	0.85886	21	0.55631	31	0.26955
2	0.99726	12	0.83298	22	0.52430	32	0.24665
3	0.99180	13	0.80559	23	0.49270	33	0.22503
4	0.98364	14	0.77690	24	0.46166	34	0.20468
5	0.97286	15	0.74710	25	0.43130	35	0.18562
6	0.95954	16	0.71640	26	0.40176	36	0.16782
7	0.94376	17	0.68499	27	0.37314	37	0.15127

8	0.92566	18	0.65309	28	0.34554	38	0.13593
9	0.90538	19	0.62088	29	0.31903	39	0.12178
10	0.88305	20	0.58856	30	0.29368	40	0.10877

Example 5.2. If we flip 8 coins, what is the probability of $B =$ "We get 3 Heads and 5 Tails"? We begin by observing that there are $2^8 = 256$ outcomes that are equally likely. A "typical" outcome in B is

$$\frac{T}{1} \frac{H}{2} \frac{H}{3} \frac{T}{4} \frac{H}{5} \frac{T}{6} \frac{T}{7} \frac{T}{8}$$

and can be described by giving the three tosses on which Heads will occur. (Here $\{2, 3, 5\}$.) So the number of outcomes in B is the number of ways of choosing three tosses for the Heads,

$$\binom{8}{3} = \frac{8 \cdot 7 \cdot 6}{1 \cdot 2 \cdot 3} = 56$$

and $P(B) = 56/256 = 0.21875$. Generalizing from the last special case we see that the probability of getting m Heads and $n - m$ Tails in n tosses is $C_{n,m}/2^n$ since there are 2^n possible outcomes for n coin tosses and there are $C_{n,m}$ ways to pick m of the n tosses for the Heads. Using the last formula to compute the probability of $H_m =$ "We get m Heads and $8 - m$ Tails" gives the following results

m	0	1	2	3	4
$C_{8,m}$	1	8	28	56	70
$P(H_m)$	.0039	.0313	.1094	.2188	.2734

We do not need to give the values for $m > 4$ since $P(H_m) = P(H_{8-m})$.

To check your understanding of the solution of the last problem, do

EXERCISE 5.1. Suppose seven coins are flipped. Compute the probability we will get (a) no Heads, (b) one Heads, (c) two Heads, (d) three Heads.

Just as the binomial coefficients can be used to compute probabilities for coin-tossing, the multinomial coefficients can be used to compute probabilities when there are $k > 2$ equally likely outcomes.

Example 5.3. To cut down on the number of possible outcomes, consider a die with 1, 2, and 3 each painted on two sides. If we roll this die 9 times, what is the probability of $A =$ "We get four 1's, two 2's, and three 3's"?

There are 3^9 outcomes that have equal probability, so it is simply a question of counting the number of outcomes in A. Extending the reasoning in Example

5.2, each outcome is specified by giving the tosses that result in 1's, 2's, and 3's, so we need to count the number of ways that $\{1, 2, \ldots, 9\}$ can be divided into three groups of sizes 4, 2, and 3. Formula (3.10) tells us that this can be done in

$$\frac{9!}{4!\,2!\,3!} = \frac{9 \cdot 8 \cdot 7 \cdot 6 \cdot 5}{1 \cdot 2 \cdot 1 \cdot 2 \cdot 3} = 1260$$

ways, so $P(A) = 1260/3^9 = 0.0640$.

Example 5.4. The problem of points is a classic problem of probability theory. Consider, for example, two groups that have each bet $48 on the outcome of a volleyball match between them. The first team to win five games wins the match, but with Team A leading 3 games to 1, a heavy rain forces them to quit playing. How should they divide up the $96 bet? Clearly, Team A should get more than half of the money, but how much more?

The general problem is: Two people or teams, which we suppose are evenly matched, are competing to see which will be the first to win n games. After A has won i games and B has won j games, the contest is interrupted and we are to decide how to divide up the prize money. This problem received a lot of attention early in the history of probability. The next table gives some of the simple recipes suggested for how much money A should receive and what they suggest in this case.

Pacioli (1494)	$i/(i+j)$	3/4
Tartaglia (1556)	$(n + i - j)/2n$	7/10
Forestani (1603)	$(2n - 1 + i - j)/(4n - 2)$	11/18

The problem of points was solved independently by Pascal and Fermat in 1654. Their idea was that A should receive a fraction of the bet equal to the probability A will win the unfinished match. To compute this probability, let $k = n - i$ be the number of games A needs to win, let $\ell = n - j$ be the number of games B needs to win, and note that the answer will depend only on (k, ℓ). For A to win the contest in exactly m games, A must win the last game and $k - 1$ of the first $m - 1$, which can be done in $C_{m-1,k-1}$ ways. Each of these possibilities has probability 2^{-m}, so summing over the possible values of m ($m \leq k + \ell - 1$ since if team B wins ℓ games they win the match), the probability A will win is

$$\sum_{m=k}^{k+\ell-1} 2^{-m} \binom{m-1}{k-1}$$

It is not surprising that no one guessed this formula. In our special case

$k = 2$ and $\ell = 4$ so the probability A will win is

$$\sum_{m=2}^{5} 2^{-m} \binom{m-1}{1} = \frac{1}{4} + \frac{2}{8} + \frac{3}{16} + \frac{4}{32}$$

$$= \frac{4+4+3+2}{16} = \frac{13}{16} = 0.8125$$

which is larger than the earlier answers (0.75 and 0.7). To check our answer, note that for B to win, the outcomes of the last five games must be $ABBBB$, $BABBB$, $BBABB$, $BBBAB$, or B can win four games in a row $BBBB$ for a total probability of $4/32 + 1/16 = 3/16$.

The next table gives the values of $p_{k,\ell} =$ the probability A wins when A needs to win k games and B needs to win ℓ. To generate this table we observed that $p_{k,1} = 1/2^k$, $p_{1,\ell} = 1 - 1/2^\ell$, and if $k, \ell > 1$ then $p_{k,\ell} = (p_{k-1,\ell} + p_{k,\ell-1})/2$.

k	$\ell = 1$	2	3	4	5	6
1	1/2	.7500	.8750	.9375	.9688	.9844
2	1/4	1/2	.6875	.8125	.8906	.9375
3	1/8	5/16	1/2	.6563	.7734	.8555
4	1/16	3/16	11/32	1/2	.6367	.7461
5	1/32	7/64	29/128	93/256	1/2	.6230
6	1/64	1/16	37/256	65/256	193/512	1/2

To explain our mixture of decimals and fractions in the table, note that $p_{i,j} = 1 - p_{j,i}$, so by looking at the appropriate half of the table (and subtracting from 1 if necessary) you can get the answer exactly or to four decimal places.

EXERCISES

5.2. If 6 balls are thrown at random into 10 boxes, what is the probability no box will contain more than 1 ball?

5.3. In a town of 50 people, one person tells a rumor to a second person, who then tells a third, and so on. If at each step the recipient of the rumor is chosen at random, what is the probability the rumor will be told 8 times without being told to someone who knows it?

5.4. An elevator in a building starts with 4 people and stops at 7 floors. If each passenger is equally likely to get off at any floor, what is the probability that no two passengers get off at the same floor?

5.5. How many people do we need so that the probability two were born on the same day of the week is more than (a) 1/3, (b) 1/2, (c) 2/3?

5.6. What is the probability that 5 people have birthdays that fall in two months but not all in the same month?

5.7. Suppose 12 coins are flipped. What is the probability we get more ($>$) Heads than Tails?

5.8. Suppose we have an urn with 6 black balls and 6 red balls. Compute the probability of getting 2 black balls and 1 red ball when we pick 3 balls out (a) without replacing them, (b) when we replace each ball after it is drawn.

5.9. **Banach's matchboxes.** An absentminded mathematician kept a box of matches in each of his two coat pockets and would pick one at random. If the two boxes started with N matches each, what is the probability that when he takes the last match out of one of the boxes, the other contains k matches?

5.10. The World Series is won by the first team to win four games. What is the probability the team that wins the first game will win the series?

5.11. Pick a number at random between 0 and 999. What is the probability (a) all three digits are different, (b) exactly two digits are the same, (c) all three are the same? (Here we regard 7 as 007.)

5.12. Pick a number at random between 0 and 9999. What is the probability (a) all four digits are different, (b) exactly two digits are the same, (c) exactly three digits are the same, (d) we get two pair, (e) all four are the same?

5.13. Suppose that we play poker by rolling five dice. How many of the 6^5 outcomes result in (a) one pair, (b) two pair, (c) three of a kind, (d) a full house, (e) four of a kind, (f) five of a kind, (g) a straight, (h) nothing?

*1.6. Probabilities of Unions

In Section 1.2, we learned that $P(A \cup B) = P(A) + P(B) - P(A \cap B)$. In this section we will extend this formula to $n > 2$ events. We begin with $n = 3$ events:

(6.1)
$$P(A \cup B \cup C) = P(A) + P(B) + P(C)$$
$$- P(A \cap B) - P(A \cap C) - P(B \cap C)$$
$$+ P(A \cap B \cap C)$$

PROOF: As in the proof of the formula for two sets, (2.4), we have to convince ourselves that the net number of times each part of $A \cup B \cup C$ is counted is 1. One way of doing this is to make a table that identifies the areas counted by each term and note that the net number of pluses in each row is 1:

	A	B	C	$A \cap B$	$A \cap C$	$B \cap C$	$A \cap B \cap C$
$A \cap B \cap C$	+	+	+	−	−	−	+
$A \cap B \cap C^c$	+	+		−			
$A \cap B^c \cap C$	+		+		−		
$A^c \cap B \cap C$		+	+			−	
$A \cap B^c \cap C^c$	+						
$A^c \cap B \cap C^c$		+					
$A^c \cap B^c \cap C$			+				

To translate this table into a proof we write an equation for each column (each equality holding because the set on the left is the disjoint union of the sets on the right):

$$P(A) = P(A \cap B \cap C) + P(A \cap B \cap C^c) + P(A \cap B^c \cap C) + P(A \cap B^c \cap C^c)$$
$$P(B) = P(A \cap B \cap C) + P(A \cap B \cap C^c) + P(A^c \cap B \cap C) + P(A^c \cap B \cap C^c)$$
$$P(C) = P(A \cap B \cap C) + P(A \cap B^c \cap C) + P(A^c \cap B \cap C) + P(A^c \cap B^c \cap C)$$
$$-P(A \cap B) = -P(A \cap B \cap C) - P(A \cap B \cap C^c)$$
$$-P(A \cap C) = -P(A \cap B \cap C) - P(A \cap B^c \cap C)$$
$$-P(B \cap C) = -P(A \cap B \cap C) - P(A^c \cap B \cap C)$$
$$P(A \cap B \cap C) = P(A \cap B \cap C)$$

Adding the seven equations gives (6.1). The last step is easy to carry out if you look at the table and recall that the sum of each row is 1.   □

The next example illustrates the use of (6.1).

Example 6.1. Suppose we roll three dice. What is the probability that we get at least one 6?

Let $A_i =$ "We get a 6 on the ith die." Clearly,

$$P(A_1) = P(A_2) = P(A_3) = 1/6$$
$$P(A_1 \cap A_2) = P(A_1 \cap A_3) = P(A_2 \cap A_3) = 1/36$$
$$P(A_1 \cap A_2 \cap A_3) = 1/216$$

So plugging into (6.1) gives

$$P(A_1 \cup A_2 \cup A_3) = 3 \cdot \frac{1}{6} - 3 \cdot \frac{1}{36} + \frac{1}{216} = \frac{108 - 18 + 1}{216} = \frac{91}{216}$$

To check this answer, we note that $(A_1 \cup A_2 \cup A_3)^c =$ "no 6" $= A_1^c \cap A_2^c \cap A_3^c$ and $|A_1^c \cap A_2^c \cap A_3^c| = 5 \cdot 5 \cdot 5 = 125$ since there are five "non-6's" that we can get on each roll. Since there are $6^3 = 216$ outcomes for rolling three dice, it follows that $P(A_1^c \cap A_2^c \cap A_3^c) = 125/216$ and $P(A_1 \cup A_2 \cup A_3) = 1 - P(A_1^c \cap A_2^c \cap A_3^c) = 91/216$.

The general formula for n events, called the **inclusion-exclusion formula**, is

$$(6.2) \quad P(\cup_{i=1}^n A_i) = \sum_{i=1}^n P(A_i) - \sum_{i<j} P(A_i \cap A_j)$$
$$+ \sum_{i<j<k} P(A_i \cap A_j \cap A_k) \ldots + (-1)^{n+1} P(A_1 \cap \ldots \cap A_n)$$

In words, we take all possible intersections of one, two, ... n sets and the signs of the sums alternate. To check the sign of the last term, note that terms with one set are $+$ and those with two are $-$, so when we have n sets, the sign is $(-1)^{n+1}$.

PROOF: Again we need to show that each area is counted once. The key to doing this is an identity that follows from the Binomial Theorem (3.5)

$$0 = (1-1)^m = \sum_{k=0}^m (-1)^k C_{m,k}$$

The term for $k = 0$ in the sum is equal to 1 so

$$(\star) \qquad 1 = -\sum_{k=1}^m (-1)^k C_{m,k} = \sum_{k=1}^m (-1)^{k+1} C_{m,k}$$

Consider now an outcome ω_0 that is in exactly m of the sets $A_1, \ldots, A_n$. The outcome ω_0 is counted $m = C_{m,1}$ times in the first sum in (6.2). In the second sum, ω_0 is counted $-C_{m,2}$ times since that is the number of ways we can pick 2 of the m sets to which ω_0 belongs. Repeating the last reasoning, we see that ω_0 is counted $C_{m,3}$ times in the third sum, ... $(-1)^{m+1} C_{m,m}$ times in the mth, and never after that. The number of times ω_0 is counted in the first m sums is exactly what appears on the right-hand side of our identity $(\star)$, so the net number of times ω_0 is counted is 1 as desired. □

Our next example is a famous application of the inclusion-exclusion formula.

Example 6.2. The Matching Problem. At a dance, n men and their n wives are paired at random. What is the probability of $B = $ "No man dances with his own wife"?

The first step in solving the problem is to describe the sample space. One possible outcome when $n = 10$ can be written as

men	1	2	3	4	5	6	7	8	9	10
women	7	10	9	4	3	8	5	1	6	2

This means that man 1 dances with woman 7, man 2 with woman 10, and so on. In general, each outcome is an arrangement of the numbers $1, \ldots, n$, with the ith number in the sequence giving the woman who danced with the ith man, so there are $n!$ possible outcomes.

Let $A_i =$ "Man i dances with his own wife." We will use the inclusion-exclusion formula to compute the probability of $\cup_{i=1}^{n} A_i =$ "At least one man dances with his own wife" and then we will subtract that answer from 1. Once the ith man is paired with his own wife, there are $(n-1)!$ ways of pairing the remaining people, so $P(A_i) = (n-1)!/n! = 1/n$. This answer should not be surprising since the ith man will choose at random from n women and one of them is his wife. Turning to $P(A_i \cap A_j)$ with $i < j$ and repeating the last argument, we see that once the ith and jth men are paired with their own wives, there are $(n-2)!$ ways of pairing the remaining people, so $P(A_i \cap A_j) = (n-2)!/n! = 1/n(n-1)$. Similar reasoning shows $|A_i \cap A_j \cap A_k| = (n-3)!$ and hence $P(A_i \cap A_j \cap A_k) = 1/n(n-1)(n-2)$.

A clear pattern has been established, so we turn to adding up the sum. There are $C_{n,m} = n(n-1)\cdots(n-m+1)/m!$ terms involving the intersection of m sets and each has the same probability so

$$P(\cup_{i=1}^{n} A_i) = n \cdot \frac{1}{n} - \frac{n(n-1)}{2} \cdot \frac{1}{n(n-1)} + \cdots + (-1)^{n+1} \cdot \frac{1}{n!}$$

$$= 1 - \frac{1}{2!} + \frac{1}{3!} \cdots + (-1)^{n+1} \frac{1}{n!}$$

To recognize the last answer as being (close to) a familiar number, we note that

$$e^{-x} = 1 - x + x^2/2! - x^3/3! + \cdots$$

so $e^{-1} = 1 - 1 + 1/2! - 1/3! + \ldots$ and

$$1 - e^{-1} = 1 - 1/2! + 1/3! - \ldots$$

This says that if n is large $P(\cup_{i=1}^{n} A_i) \approx 1 - e^{-1}$, and hence $P(B) \approx e^{-1}$. To get a bound on the error in the approximation, we note that the series for e^{-1} is an alternating series with decreasing terms, so the successive partial sums give upper and lower bounds on the limit, and hence

(6.3) $$|P(B) - e^{-1}| \leq 1/(n+1)!$$

The last inequality shows that $1/e$ is a very good approximation even when n is small. For example, when $n = 4$ our bound shows the error is at most $1/5! = 1/120 = 0.0083$, while an exact solution (see Exercise 6.4) shows $P(B) = 9/24 = 0.3750$, which is 0.0071 larger than $1/e = 0.3679$.

Our final computation concerning the matching problem anticipates the Poisson distribution, to be introduced in Section 3.1. Let N_n be the number of men who dance with their own wives, with the subscript n indicating the total number of men. Having worked hard to compute $P(N_n = 0)$, we can now easily compute $P(N_n = k)$ for $k \geq 1$. To do this, let $a_n = n!P(N_n = 0)$ be the number of pairings in which no man dances with his own wife and observe that the number of pairings in which exactly k men dance with their own wives is $\binom{n}{k}a_{n-k}$, i.e., the number of ways of picking out the k men who dance with their own wives times the number of ways that the remaining $n - k$ men can be paired so that none of them dances with his own wife. From the last observation it follows that

$$P(N_n = k) = \frac{1}{n!}\binom{n}{k}a_{n-k}$$

$$= \frac{1}{k!}\frac{a_{n-k}}{(n-k)!} = \frac{1}{k!}P(N_{n-k} = 0) \approx e^{-1}\frac{1}{k!}$$

The Bonferroni Inequalities. By doing the proof of (6.2) more carefully, one can conclude that

(6.4) $\quad P(\cup_{i=1}^{n} A_i) \leq \sum_{i=1}^{n} P(A_i)$

(6.5) $\quad P(\cup_{i=1}^{n} A_i) \geq \sum_{i=1}^{n} P(A_i) - \sum_{i<j} P(A_i \cap A_j)$

(6.6) $\quad P(\cup_{i=1}^{n} A_i) \leq \sum_{i=1}^{n} P(A_i) - \sum_{i<j} P(A_i \cap A_j) + \sum_{i<j<k} P(A_i \cap A_j \cap A_k)$

and so on. In brief, if you stop the inclusion-exclusion formula with a + term you get an upper bound; if you stop with a − term you get a lower bound. The first inequality was proved in Section 1.2. (See (2.5).) Proofs of the second and third are sketched in Exercises 6.15 and 6.16. The next two examples illustrate the use of these inequalities.

Example 6.3. Suppose we roll a die 15 times. What is the probability that we do not see all 6 numbers at least once?

Let A_i be the event that we never see i. $P(A_i) = 5^{15}/6^{15}$ since there are 6^{15} outcomes in all but only 5^{15} that contain no i's. $5^{15}/6^{15} = 0.064905$, so the first bound in (6.4) gives us

$$P\left(\cup_{i=1}^{6} A_i\right) \le 6(0.064905) = 0.389433$$

Turning to the second, we note that for any $i < j$, we have $P(A_i \cap A_j) = 4^{15}/6^{15} = 0.002284$ and there are $C_{6,2} = (6 \cdot 5)/2 = 15$ choices for $i < j$ so

$$P\left(\cup_{i=1}^{6} A_i\right) \ge 0.389433 - 15(0.002284) = 0.355178$$

To compute the third bound we note that for any $i < j < k$, we have $P(A_i \cap A_j \cap A_k) = 3^{15}/6^{15} = 3.05 \times 10^{-5}$ and there are $C_{6,3} = (6 \cdot 5 \cdot 4)/3! = 20$ choices for $i < j < k$ so

$$P\left(\cup_{i=1}^{6} A_i\right) \le 0.355178 + 20(3.05 \times 10^{-5}) = 0.355788$$

This answer is already quite accurate, but one can continue to find the exact answer:

$$C_{6,1}(5/6)^{15} - C_{6,2}(4/6)^{15} + C_{6,3}(3/6)^{15} - C_{6,4}(2/6)^{15} + C_{6,5}(1/6)^{15}$$
$$= 0.355788 - 1.045 \times 10^{-6} + 1.276 \times 10^{-11} = 0.355787$$

Notice that the first inequality gives a crude but useful upper bound while the second inequality is quite close to the answer.

Example 6.4. Returning to the Birthday Problem, suppose there are 15 people in the class. What is the probability two people have the same birthday?

For $i \ne j$ let $A_{\{i,j\}}$ be the event that i and j have the same birthday. $P(A_{\{i,j\}}) = 1/365$ and there are $(15 \cdot 14)/2 = 105$ choices for the people so

$$P(\cup_{\{i,j\}} A_{\{i,j\}}) \le 105/365 = 0.28767$$

To compute a lower bound we have to subtract the probability of the intersection of all pairs of sets, that is,

$$\frac{1}{2} \sum_{\{i,j\} \ne \{k,\ell\}} P(A_{\{i,j\}} \cap A_{\{k,\ell\}})$$

We need the 1/2 since each term appears twice. To evaluate the sum, we observe $P(A_{\{i,j\}} \cap A_{\{k,\ell\}}) = 1/365^2$. (There are two cases to consider: $\{i,j\}$ and $\{k,\ell\}$ overlap or are disjoint.) So our lower bound is

$$0.28767 - \frac{105 \cdot 104}{2} \frac{1}{365^2} = 0.24668$$

The computation of the third bound is complicated by the fact that

$$P(A_{\{i,j\}} \cap A_{\{j,k\}} \cap A_{\{k,i\}}) = 1/365^2$$

but other triple intersections have probability $1/(365)^3$. Since the choice of i, j, k determines the three sets $\{i, j\}$, $\{j, k\}$, and $\{k, i\}$, the third sum is

$$\binom{15}{3} \frac{1}{(365)^2} + \left\{ \binom{105}{3} - \binom{15}{3} \right\} \frac{1}{(365)^3}$$

$$= \frac{105 \cdot 104 \cdot 103}{3! \cdot (365)^3} + \frac{15 \cdot 14 \cdot 13}{3!} \left\{ \frac{1}{(365)^2} - \frac{1}{(365)^3} \right\}$$

$$= 0.003855 + 0.003405 = 0.007260$$

Adding this to the second bound 0.24668 gives 0.25394, which is very close to the exact answer from the table in Section 1.5: 0.2529.

Example 6.5. In the summer of 1941, Joe DiMaggio had what many people consider the greatest record in sports, the "streak" in which he had at least one hit in each of 56 games. To compute the probability of this event we will introduce three assumptions that are somewhat questionable: (i) a player gets exactly four at bats per game (during the streak, DiMaggio averaged 3.98 at bats per game), (ii) the outcomes of different at bats are independent with the probability of a hit being 0.325, Joe DiMaggio's lifetime batting average, and (iii) the outcomes for successive games are independent. From assumptions (i) and (ii) it follows that the probability p of a hit during a game is $1 - (0.675)^4 = 0.7924$. We leave it to the reader to generalize our model to allow for the number of at bats in a game to be random or for the probability of a hit to depend on the number of previous failures in the same game. With such generalizations in mind, we will express the probability of a streak in terms of the probability p of getting at least one hit in a game.

Assuming a 162-game season, we could let A_i be the probability that a player got hits in games $i, i+1, \ldots i+55$ for $1 \leq i \leq 107$. However, these events have a problem: if A_i occurs, it becomes much easier for A_{i+1}, A_{i-1}, and other "nearby" events to occur. To avoid this problem, we will let B_0 be the event that the player gets hits in games $1, 2, \ldots, 56$, and for $1 \leq i \leq 106$ let B_i be the event that the player got no hit in game i but got hits in games $i+1, i+2, \ldots, i+56$. With this scheme we have to treat B_0 separately, but this headache is more than made up for by the fact that $B_i \cap B_j = \emptyset$ if $0 \leq i < j \leq i + 56$. (B_i needs a hit in game j but B_j must have no hit then.)

The event of interest $S = \cup_{i=0}^{106} B_i$ so

$$P(S) \leq \sum_{i=0}^{106} P(B_i) = p^{56} + 106(1-p)p^{56} = 5.045 \times 10^{-5}$$

for our choice of p. To compute the second bound we begin by noting that $P(B_0 \cap B_j) = p^{56}(1-p)p^{56}$ when $56 < j \leq 106$, and that $P(B_i \cap B_j) = (1-p)p^{56}(1-p)p^{56}$ when $i > 0$ and $56 + i < j \leq 106$. There are 50 terms of the first type and $49 + 48 + \cdots + 2 + 1 = (50 \cdot 49)/2 = 1225$ terms of the second type so

$$\sum_{0 \leq i < j \leq 106} P(B_i \cap B_j) = 50(1-p)p^{112} + 1225(1-p)^2 p^{112} = 3.036 \times 10^{-10}$$

Since this is the number we have to subtract from the upper bound to get the lower bound, our upper bound is extremely accurate.

REMARK. We would not have done this well if we had used the A's instead. The corresponding upper bound is

$$P(S) \leq \sum_{i=1}^{107} P(A_i) = 107 p^{56} = 2.346 \times 10^{-4}$$

for the value of p we are considering. Comparing formulas, we see that this is about $1/(1-p) \approx 5$ times the upper bound from the B's. The first bound is bad but the second turns out to be a disaster.

$$\sum_{1 \leq i < j \leq 107} P(A_i \cap A_j) = p^{56} \sum_{i=1}^{106} \sum_{j=i+1}^{107} p^{\min(j-i,56)}$$
$$= p^{56} \left(106p + 105p^2 + \cdots 51p^{55} + 1326p^{56} \right) = 8.554 \times 10^{-4}$$

for our value of p. Subtracting this from the first bound, our second bound is -6.208×10^{-4}, which is < 0.

The next exercise gives another example where the Bonferroni inequalities do not work very well.

EXERCISE 6.1. Use (6.4) and (6.5) to compute an upper and a lower bound on the probability that in a group of 30 people, at least two were born on the same day.

EXERCISES

6.2. In a certain city 60% of the people subscribe to newspaper A, 50% to B, 40% to C, 30% to A and B, 20% to B and C, and 10% to A and C, but no one subscribes to all three. What percentage subscribe to (a) at least one newspaper, (b) exactly one newspaper?

6.3. Santa Claus has 45 drums, 50 cars, and 55 baseball bats in his sled. 15 boys will get a drum and a car, 20 a drum and a bat, 25 a bat and a car, and 5 will get three presents. (a) How many boys will receive presents? (b) How many boys will get just a drum?

6.4. Four men put their car keys into a hat and then each picks one out at random. Let N be the number of men who get their own keys back. Compute the probability $N = 4, 2, 1, 0$.

6.5. Compute the probability of no matches in the Matching Problem when $n = 6$. To do this it is convenient to "decompose the permutation into cycles." To explain what this means, consider the example

men	1	2	3	4	5	6
women	3	4	5	2	6	1

In this case decomposition is (1356)(24), indicating that 1 danced with 3's wife, 3 with 5's wife, 5 with 6's wife, 6 with 1's wife, 2 with 4's wife, and 4 with 2's wife. Compute the probability of no matches by considering the size of the cycle 1 is contained in.

6.6. The approximation in (6.3) is so accurate that we can use it to determine the exact answer. $P(B)$ must have the form $m/n!$ and be within $1/(n+1)!$ of $1/e$, but for $n \geq 1$ there is only one such value of m. Use this observation to compute the exact value of the probability for $n = 8$.

6.7. Ten people call an electrician and ask him to come to their houses on randomly chosen days of the work week (Monday through Friday). What is the probability of $A =$ "He has at least one day with no jobs"?

6.8. We pick a number between 0 and 999, then a computer picks one at random from that range. Use (6.1) to compute the probability at least two of our digits will match the computer's number. (Note: We include any leading zeros, so 017 and 057 have two matching digits.)

6.9. You pick 13 cards out of a deck of 52. What is the probability that you will not get a card from every suit?

6.10. You pick 13 cards out of a deck of 52. Let $A =$ "You have exactly six cards in at least one suit" and $B =$ "You have exactly six spades." (6.4) says that $P(A) \leq 4P(B)$. Compute $P(A)$ and $P(A)/P(B)$.

6.11. Suppose we roll seven four-sided dice. Use (6.4) and (6.5) to give upper and lower bounds on the probability we have at least five of the same number.

6.12. Suppose we roll two dice 6 times. Use (6.4)–(6.6) to compute bounds on the probability of $A =$ "We get at least one double 6." Compare the bounds with the exact answer $1 - (35/36)^6$.

6.13. Use (6.4) and (6.5) to compute an upper and a lower bound on the probability that in a group of 60 people, at least 3 were born on the same day.

6.14. Suppose we try 20 times for an event with probability 0.01. Use (6.4)–(6.6) to compute bounds on the probability of one success.

6.15. Prove (6.5) by noting that an outcome that is in m of the sets is counted m times in the first term and $-m(m-1)/2$ times in the second.

6.16. Prove (6.6) by noting that an outcome that is in m of the sets is counted m times in the first sum, $-m(m-1)/2$ times in the second, and $m(m-1)(m-2)/6$ times in the third.

6.17. Suppose we put n balls at random into r boxes. Let $p_k(r, n)$ be the probability that there are k empty boxes. Show that

$$p_0(r, n) = \sum_{j=0}^{n} (-1)^r \binom{r}{j} \left(1 - \frac{j}{r}\right)^n$$

$$p_k(r, n) = \binom{r}{k} \left(1 - \frac{k}{n}\right)^n p_0(r - k, n)$$

This may sound like a carnival game but this question comes up in a number of different disguises. Suppose there are r baseball cards in a set, and we have n. Then $p_k(r, n)$ gives the probability we are missing exactly k of them.

6.18. Let $\Delta_{i=1}^{n} A_i$ be the outcomes that are in exactly one of the n sets. (a) Find a formula like (6.1) for $P(\Delta_{i=1}^{3} A_i)$. (b) Find a formula like (6.2) for $P(\Delta_{i=1}^{n} A_i)$.

1.7. Chapter Summary and Review Problems

In this section, as we will do at the end of each chapter, we summarize the main results of this chapter. Formula numbers here indicate where the facts can be found in the text.

Section 1.1 Experiments and events. The sample space gives the set of possible **outcomes** of an **experiment**. An **event** is a subset of the sample space. In Section 1.1 the basic definitions of set theory were reviewed.

Section 1.2. A **probability** is way of assigning numbers to events that satisfies

(i) For any event A, $0 \le P(A) \le 1$.

(ii) If Ω is the sample space, $P(\Omega) = 1$.

(iii) For a finite or infinite sequence of disjoint events

$$P(\cup_i A_i) = \sum_i P(A_i)$$

These assumptions are motivated by the **frequency interpretation of probability**, which says that if we repeat an experiment a large number of times then the fraction of times the event A occurs will be close to $P(A)$. Some simple consequences of the definition are

(2.1) $P(A^c) = 1 - P(A)$

(2.2) $P(\emptyset) = 0$

(2.3) If $A \subset B$ then $P(A) \leq P(B)$

(2.4) $P(A \cup B) = P(A) + P(B) - P(A \cap B)$

(2.5) $P(\cup_i A_i) \leq \sum_i P(A_i)$

Section 1.3. Permutations and combinations. In many situations, $P(A)$ is the fraction of outcomes that lie in A. To compute probabilities in these situations we have to be able to count the number of outcomes. To do this the following results are useful:

(3.1) **The multiplication rule.** Suppose that m experiments are performed in order and that, no matter what the outcomes of experiments $1, \ldots, k-1$ are, experiment k has n_k possible outcomes. Then the total number of outcomes is $n_1 \cdot n_2 \cdots n_k$.

Permutations. If we have n people then the number of ways we can pick k of them to stand in line is the number of **permutations of n things taken k at a time**

(3.3) $$P_{n,k} = n \cdot (n-1) \cdots (n-k+1) = \frac{n!}{(n-k)!}$$

Here, n **factorial**, $n! = n \cdot (n-1) \cdots 2 \cdot 1$.

Combinations. If we have n people then the number of ways we can pick a set of k from them is the number of **combinations of n things taken k at a time**

(3.4) $$C_{n,k} = \frac{n \cdot (n-1) \cdots (n-k+1)}{k!} = \frac{n!}{k!(n-k)!} = \binom{n}{k}$$

The **binomial theorem** says

(3.5) $$(a+b)^n = \sum_{k=0}^{n} \binom{n}{k} a^k b^{n-k}$$

where we have used the convention that $0! = 1$ so $\binom{n}{0} = \binom{n}{n} = 1$.

Multinomial coefficients. The number of ways we can divide n things into m piles of sizes $n_1, \ldots, n_m$ is

(3.10)
$$\frac{n!}{n_1! n_2! \cdots n_m!}$$

Section 1.4. Urn problems. Suppose we have an urn with M red balls and N black balls and we draw out n balls without replacement. Then the probability we get r red balls and hence $n - r$ black balls is

$$\frac{\binom{M}{r}\binom{N}{n-r}}{\binom{M+N}{n}}$$

where we have used the convention that $\binom{m}{j} = 0$ if $j < 0$ or $j > m$. The last answer generalizes in a straightforward way to situations with more than two colors.

Section 1.5. Repeated experiments. The problems in this section concern drawing with replacement from an urn with K balls numbered from 1 up to K. That is, we pick a ball, note its number and then return it to the urn. Since all K outcomes are possible on each trial the total number of outcomes is K^k. Two important special cases are:

The Birthday problem. If we draw k balls with replacement, then the probability all the numbers are different is

$$\frac{P_{K,k}}{K^k} = \frac{K}{K} \cdot \frac{(K-1)}{K} \cdots \frac{(K-k+1)}{K}$$

Flipping coins. This corresponds to $K = 2$, $1 = $ Heads, $2 = $ Tails. The probability of exactly j Heads in k trials is

$$\binom{k}{j} 2^{-k}$$

The first factor gives the number of ways of picking j tosses on which Heads occurs; the second, the probability of each of the 2^k outcomes.

Section 1.6. The inclusion-exclusion formula says

(6.2) $P(\cup_{i=1}^n A_i) = \displaystyle\sum_{i=1}^n P(A_i) - \sum_{i<j} P(A_i \cap A_j)$

$$+ \sum_{i<j<k} P(A_i \cap A_j \cap A_k) \ldots + (-1)^{n+1} P(A_1 \cap \ldots \cap A_n)$$

In words, we take all possible intersections of one, two, ... n sets and the signs of the sums alternate. To check the sign of the last term, note that terms with one set are $+$ and those with two are $-$, so when we have n sets, the sign is $(-1)^{n+1}$.

The **Bonferroni inequalities** say

$$(6.4) \quad P(\cup_{i=1}^{n} A_i) \leq \sum_{i=1}^{n} P(A_i)$$

$$(6.5) \quad P(\cup_{i=1}^{n} A_i) \geq \sum_{i=1}^{n} P(A_i) - \sum_{i<j} P(A_i \cap A_j)$$

$$(6.6) \quad P(\cup_{i=1}^{n} A_i) \leq \sum_{i=1}^{n} P(A_i) - \sum_{i<j} P(A_i \cap A_j) + \sum_{i<j<k} P(A_i \cap A_j \cap A_k)$$

and so on. In brief, if you stop the inclusion-exclusion formula with a $+$ term you get an upper bound; if you stop with a $-$ term you get a lower bound.

REVIEW PROBLEMS

7.1. A tire manufacturer wants to test four different types of tires on three different types of roads at five different speeds. How many tests are required?

7.2. A tourist wants to visit six of America's ten largest cities. In how many ways can she do this if the order of her visits is (a) important, (b) not important?

7.3. How many ways can four rooks be put on a chessboard so that no rook can capture any other rook? Or, what is the same: How many ways can eight markers be placed on an 8 × 8 grid of squares so that there is at most one in each row or column?

7.4. Four men and four women are shipwrecked on a tropical island. How many ways can they (a) form four male-female couples, (b) get married if we keep track of the order in which the weddings occur, (c) divide themselves into four unnumbered pairs, (d) split up into four groups of two to search the North, East, West, and South shores of the island, (e) walk single-file up the ramp to the ship when they are rescued, (f) take a picture to remember their ordeal if all eight stand in a line but each man stands next to his wife?

7.5. Six married couples are seated at a round table. Compute the probability that some wife sits next to her husband.

7.6. How many four-letter "words" can you make if no letter is used twice and each word must contain at least one vowel (A, E, I, O or U)?

7.7. Assuming all phone numbers are equally likely, what is the probability that all the numbers in a seven-digit phone number are different?

7.8. In the freshman class, 62% of the sudents take math, 49% take science, and 38% take both science and math. What percentage takes at least one science or math course?

7.9. Show that $P(A \cap B \cap C) \geq P(A) + P(B) + P(C) - 2P(A \cup B \cup C)$. When will equality hold?

7.10. Suppose $P(A) = 1/3$, $P(A^c \cap B^c) = 1/2$, and $P(A \cap B) = 1/4$. What is $P(B)$?

7.11. Use the inclusion-exclusion formula to compute the probability that a randomly chosen number between 0000 and 9999 contains at least one 1.

7.12. Three fair dice are rolled. What is the probability the sum is 14?

7.13. In a dice game the "dealer" rolls two dice, the player rolls two dice, and the player wins if his total is larger ($>$) than the dealer's. (a) What is the probability the player wins? (b) Would the player's chances be better or worse if three dice were used?

7.14. In a kindergarten class of 20 students, one child is picked each day to help serve the morning snack. What is the probability that in one week five different children are chosen?

7.15. Four red cards (i.e., hearts and diamonds) and four black cards are face down on the table. A psychic who claims to be able to locate the four black cards turns over 4 cards and gets 3 black cards and 1 red card. What is the probability he would do this if he were guessing?

7.16. An investor picks 3 stocks out of 10 recommended by his broker. Of these, six will show a profit in the next year. What is the probability the investor will pick (a) 3 (b) 2 (c) 1 (d) 0 profitable stocks?

7.17. A small settlement contains five families each consisting of four people. If 6 of these 20 individuals, whom we assume to be chosen at random, have a contagious disease, what is the probability that (a) only 2 families are affected, (b) only 1 family has no sick individuals, (c) all 4 families have someone sick?

7.18. Two balls are drawn from an urn with balls numbered from 1 up to 10. What is the probability that the two numbers will differ by more ($>$) than three?

7.19. A town council considers the question of closing down an "adult" theatre. The five men on the council all vote against this and the three women vote in favor. What is the probability we would get this result (a) if the council members determined their votes by flipping a coin? (b) if we assigned the five "no" votes to council members chosen at random?

7.20. An urn contains white balls numbered 1 to 15 and black balls also numbered 1 to 15. Suppose you draw 4 balls. What is the probability that (a) no two have the same number? (b) you get exactly one pair with the same number? (c) you get two pair with the same numbers?

7.21. An urn contains five red balls and five black balls. Players A and B draw balls from this urn alternately until one person draws a red ball and wins the game. What is the probability A, who draws first, will win the game?

7.22. Two evenly matched players (A and B) play a game until one has won three times. What is the probability the winner was never behind in the match? For example $ABBAA$ is not possible because after the third game A was behind 2 games to 1, but $ABABA$ is a possible outcome.

7.23. In seven-card stud you receive seven cards and use them to make the best poker hand you can. Ignoring the possibility of a straight or a flush compute the probability that you have (a) four of a kind, (b) a full house, (c) three of a kind, (d) two pair, and (e) one pair.

7.24. Suppose you draw seven cards out of a deck of 52. What is the probability you will have at least five cards of one suit?

2 Conditional Probability

2.1. Independence

Intuitively, two events A and B are independent if the occurrence of A has no influence on the probability of occurrence of B. The formal definition is: A and B are **independent** if $P(A \cap B) = P(A)P(B)$. We now give three classic examples of independent events. In each case it should be clear that the intuitive definition is satisfied, so we will only check the conditions of the formal one.

Example 1.1. Flip two coins. $A = $ "The first coin shows Heads," $B = $ "The second coin shows Heads." $P(A) = 1/2$, $P(B) = 1/2$, $P(A \cap B) = 1/4$.

Example 1.2. Roll two dice. $A = $ "The first die shows 5," $B = $ "The second die shows 2." $P(A) = 1/6$, $P(B) = 1/6$, $P(A \cap B) = 1/36$.

Example 1.3. Pick a card from a deck of 52. $A = $ "The card is an ace," $B = $ "The card is a spade." $P(A) = 1/13$, $P(B) = 1/4$, $P(A \cap B) = 1/52$.

Two examples of events that are not independent are

Example 1.4. Draw two cards from a deck. $A = $ "The first card is a spade," $B = $ "The second card is a spade." $P(A) = 1/4$, $P(B) = 1/4$, but

$$P(A \cap B) = \frac{C_{13,2}}{C_{52,2}} = \frac{13 \cdot 12}{52 \cdot 51} < \left(\frac{1}{4}\right)^2$$

Intuitively, these two events are not independent, since getting a spade the first time reduces the fraction of spades in the deck and makes it harder to get a spade the second time. Anticipating a result in the next section, note that we have a probability of 13/52 of getting a spade the first time and, if we succeed, only a 12/51 chance the second time.

Example 1.5. Roll two dice. $A =$ "The sum of the two dice is 9," $B =$ "The first die is 2." $A = \{(6,3), (5,4), (4,5), (3,6)\}$, so $P(A) = 4/36$. $P(B) = 1/6$, but $P(A \cap B) = 0$ since $(2,7)$ is impossible. In general if A and B are disjoint events that have positive probability, they are not independent since $P(A)P(B) > 0 = P(A \cap B)$.

EXERCISE 1.1. Roll two dice. Let $A =$ "The sum of the two dice is 7," $B =$ "The first die is 4," $C =$ "The difference of the two dice is even." (a) Are A and B independent? (b) B and C? (c) C and A?

There are two ways of extending the definition of independence to more than two events. $A_1, \ldots, A_n$ are said to be **pairwise independent** if for each $i \neq j$, $P(A_i \cap A_j) = P(A_i)P(A_j)$, that is, each pair is independent. $A_1, \ldots, A_n$ are said to be **independent** if for any $1 \leq i_1 < i_2 < \ldots < i_k \leq n$ we have

$$P(A_{i_1} \cap \ldots \cap A_{i_k}) = P(A_{i_1}) \cdots P(A_{i_k})$$

If we flip n coins and let $A_i =$ "The ith coin shows Heads," then the A_i are independent since $P(A_i) = 1/2$ and $P(A_{i_1} \cap \ldots \cap A_{i_k}) = 1/2^k$. We have already seen an example of events that are pairwise independent but not independent:

Example 1.6. Birthdays. Let $A =$ "Alice and Betty have the same birthday" $B =$ "Betty and Carol have the same birthday," $C =$ "Carol and Alice have the same birthday." Since there are 365 ways two girls can have the same birthday out of 365^2 possibilities (as in Example 5.1 of Chapter 1, we are assuming that leap year does not exist and that all the birthdays are equally likely), $P(A) = P(B) = P(C) = 1/365$. Likewise, there are 365 ways all three girls can have the same birthday out of 365^3 possibilities, so

$$P(A \cap B) = \frac{1}{365^2} = P(A)P(B)$$

i.e., A and B are independent. Similarly, B and C, are independent and C and A are independent, so A, B, and C are pairwise independent. The three events A, B, and C are not independent, however, since $A \cap B = A \cap B \cap C$ and hence

$$P(A \cap B \cap C) = \frac{1}{365^2} \neq \left(\frac{1}{365}\right)^3 = P(A)P(B)P(C)$$

The last example is somewhat unusual. However, the moral of the story is that to show several events are independent, you have to check more than just that each pair is independent.

Example 1.7. Suppose we roll 6 dice. What is the probability of $A =$ "We get exactly two 4's"?

One way that A can occur is

$$\frac{\times\ 4\ \times\ 4\ \times\ \times}{1\ 2\ 3\ 4\ 5\ 6}$$

where $\times$ stands for "not a 4." Since the six events "die one shows $\times$," "die two shows 4," ..., "die six shows $\times$" are independent, the indicated pattern has probability

$$\frac{5}{6}\cdot\frac{1}{6}\cdot\frac{5}{6}\cdot\frac{1}{6}\cdot\frac{5}{6}\cdot\frac{5}{6} = \left(\frac{1}{6}\right)^2\left(\frac{5}{6}\right)^4$$

Here we have been careful to say "pattern" rather than "outcome" since the given pattern corresponds to 5^4 outcomes in the sample space of 6^6 possible outcomes for 6 dice. Each pattern that results in A corresponds to a choice of 2 of the 6 trials on which a 4 will occur, so the number of patterns is $C_{6,2}$. When we write out the probability of each pattern there will be two 1/6's and four 5/6's so each pattern has the same probability and

$$P(A) = \binom{6}{2}\left(\frac{1}{6}\right)^2\left(\frac{5}{6}\right)^4$$

Example 1.8. The binomial distribution. Generalizing from the last example, suppose we perform an experiment n times and on each trial an event we call "success" has probability p. (Here and in what follows, when we repeat an experiment, we assume that the outcomes of the various trials are independent.) Then the probability of k successes is

(1.1)
$$\binom{n}{k}p^k(1-p)^{n-k}$$

Taking $n = 6$, $k = 2$, and $p = 1/6$ in (1.1) gives the answer in the previous example. The reasoning for the general formula is similar. There are $C_{n,k}$ ways of picking k of the n trials for successes to occur, and each pattern of k successes and $n - k$ failures has probability $p^k(1-p)^{n-k}$.

EXERCISE 1.2. A student takes a test with 10 multiple-choice questions. Since she has never been to class she has to choose at random from the 4 possible answers. What is the probability she will get exactly 3 right?

EXERCISE 1.3. A football team wins each week with probability 0.7 and loses with probability 0.3. If we suppose that the outcomes of their 10 games are independent, what is the probability they will win exactly 8 games?

EXERCISE 1.4. A die is rolled 8 times. What is the probability we will get exactly two 3's?

The arguments above generalize easily to independent events with more than two possible outcomes.

Example 1.9. The multinomial distribution. Consider a die with 1 painted on three sides, 2 painted on two sides, and 3 painted on one side. If we roll this die ten times what is the probability we get five 1's, three 2's and two 3's?

The answer is

$$\frac{10!}{5!\,3!\,2!} \left(\frac{1}{2}\right)^5 \left(\frac{1}{3}\right)^3 \left(\frac{1}{6}\right)^2$$

The first factor (by (3.10) in Chapter 1) gives the number of ways to pick five rolls for 1's, three rolls for 2's, and two rolls for 3's. The second factor gives the probability of any outcome with five 1's, three 2's, and two 3's. Generalizing from this example, we see that if we have k possible outcomes for our experiment with probabilities $p_1, \ldots, p_k$ then the probability of getting exactly n_i outcomes of type i in $n = n_1 + \cdots + n_k$ trials is

(1.2)
$$\frac{n!}{n_1! \cdots n_k!} \, p_1^{n_1} \cdots p_k^{n_k}$$

since the first factor gives the number of outcomes and the second the probability of each one.

EXERCISE 1.5. A baseball player gets a hit with probability 0.3, a walk with probability 0.1, and an out with probability 0.6. If he bats 4 times during a game and we assume that the outcomes are independent, what is the probability he will get 1 hit, 1 walk, and 2 outs?

EXERCISE 1.6. The output of a machine is graded excellent 70% of the time, good 20% of the time, and defective 10% of the time. What is the probability a sample of size 15 has 10 excellent, 3 good, and 2 defective items?

The binomial (or, more generally, multinomial) distribution comes up in a number of situations. Another commonly occurring distribution associated with independent events is

Example 1.10. The geometric distribution. Suppose we roll a die repeatedly until a 6 occurs, and let N be the number of times we roll the die. Using

× to denote "not a 6," we have

$$P(N = 1) = P(6) = \frac{1}{6}$$

$$P(N = 2) = P(\times 6) = \frac{5}{6} \cdot \frac{1}{6}$$

$$P(N = 3) = P(\times \times 6) = \frac{5}{6} \cdot \frac{5}{6} \cdot \frac{1}{6}$$

From the first three terms it is easy to see that for $k \geq 1$

$$P(N = k) = P(\times \text{ on } k - 1 \text{ rolls then } 6) = \left(\frac{5}{6}\right)^{k-1} \frac{1}{6}$$

Generalizing, we see that if we are waiting for an event of probability p, the number of trials needed, N, has

(1.3) $$P(N = k) = (1 - p)^{k-1}p \quad \text{for } k = 1, 2, \ldots$$

since $\{N = k\}$ occurs exactly when we have $k - 1$ failures followed by a success.

Example 1.11. The negative binomial distribution. Suppose we repeat an experiment with a probability p of success until we have n successes and let T_n be the number of trials required. If $n = 4$, one possible realization is

$$\frac{F\ S\ F\ F\ F\ S\ S\ F\ F\ S}{1\ 2\ 3\ 4\ 5\ 6\ 7\ 8\ 9\ 10}$$

Now $P(T_4 = 10) = C_{9,6}\, p^4(1 - p)^6$ since (i) any outcome in $\{T_4 = 10\}$ will have 4 successes and 6 failures and hence have probability $p^4(1 - p)^6$, and (ii) there must be a success on the 10th trial, so the number of such outcomes is the number of ways 6 failures can occur in the first 9 trials, i.e., $C_{9,6}$. Generalizing, we see that

(1.4) $$P(T_n = n + k) = C_{n+k-1,k}\, p^n(1 - p)^k \quad \text{for } k \geq 0$$

since (i) any outcome in $\{T_n = n + k\}$ will have n successes and k failures and hence will have probability $p^n(1 - p)^k$, and (ii) there must be a success on the $(n + k)$th trial, so the number of such outcomes is the number of ways k failures can occur in the first $n + k - 1$ trials, i.e., $C_{n+k-1,k}$.

The name "negative binomial distribution" comes from the fact that if we define

$$\binom{x}{k} = \frac{x(x - 1) \cdots (x - k + 1)}{k!}$$

for any real number x then for $k \geq 0$

$$P(T_n = n + k) = \frac{(n + k - 1)!}{k!(n-1)!} p^n (1-p)^k$$

$$= \frac{n(n+1)\cdots(n+k-1)}{k!} p^n (1-p)^k$$

$$= \frac{-n(-n-1)\cdots(-n-(k-1))}{k!} p^n \{-(1-p)\}^k$$

$$= \binom{-n}{k} p^n \{-(1-p)\}^k$$

The extension of the definition of $\binom{x}{k}$ given above is useful for more than just rewriting the definition of the negative binomial. For any x and $-1 < t < 1$ we have **Newton's binomial formula,**

(1.5)
$$(1+t)^x = \sum_{k=0}^{\infty} \binom{x}{k} t^k$$

where $\binom{x}{0} = 1$. If x is an integer then the coefficients of t^{x+j} are 0 for $j \geq 1$ and this reduces to the binomial theorem ((3.5) in Chapter 1). To see why (1.5) is true, recall that (under suitable assumptions)

$$f(t) = f(0) + \sum_{k=1}^{\infty} f^{(k)}(0) \frac{t^k}{k!}$$

where $f^{(k)}$ is the kth derivative of f. When $f(t) = (1+t)^x$, $f^{(k)}(0)/k! = \binom{x}{k}$.

EXERCISES

1.7. Suppose we draw two cards out of a deck of 52. Let $A = $ " The first card is an Ace," $B = $ "The second card is a spade." Are A and B independent?

1.8. A family has three children, each of whom is a boy or a girl with probability $1/2$. Let $A = $ "There is at most 1 girl," $B = $ "The family has children of both sexes." (a) Are A and B independent? (b) Are A and B independent if the family has four children?

1.9. Suppose we roll a red and a green die. Let $A = $ "The red die shows a 2 or a 5," $B = $ "The sum of the two dice is at least 7." Are A and B independent?

1.10. Consider n flips of a fair coin. For $k < n$, let A_k be the event that a run is completed at time k, $= $ "The results of the kth and $(k+1)$th flips are different." For example, if $n = 10$ and the outcomes of the first 10 flips are

HHHTTHHTTH, then runs are completed at times 3, 5, 7, and 9. Show that the events A_k, $1 \le k < n$ are independent.

1.11. Roll two dice. Let A = "The first die is odd," B = "The second die is odd," and C = "The sum is odd." Show that these events are pairwise independent but not independent.

1.12. Show that A and B^c are independent if A and B are.

1.13. Give an example to show that we may have (i) A and B are independent, (ii) A and C are independent, but (iii) A and $B \cup C$ are not independent. Can you do this if B and C are disjoint?

1.14. Show that if A, B, and C are independent then (a) A and $B \cap C$ are, (b) A and $B \cup C$ are.

1.15. Two students, Alice and Betty, are registered for a statistics class. Alice attends 80% of the time, Betty 60% of the time, and their absences are independent. On a given day, what is the probability (a) at least one of these students is in class (b) exactly one of them is there?

1.16. Let A and B be two independent events with $P(A) = 0.4$ and $P(A \cup B) = 0.64$. What is $P(B)$?

1.17. Suppose $A_1, \ldots, A_n$ are independent. Show that

$$P(A_1 \cup \cdots \cup A_n) = 1 - \left\{ \prod_{m=1}^{n} (1 - P(A_m)) \right\}$$

1.18. Three students each have probability 1/3 of solving a problem. What is the probability at least one of them will solve the problem?

1.19. Three couples that were invited to dinner will independently show up with probabilities 0.9, 8/9, and 0.75. Let N be the number of couples that show up. Calculate the probability $N = 3, 2, 1, 0$.

1.20. A baseball player is said to "hit for the cycle" if he has a single, a double, a triple, and a home run all in one game. Suppose these four types of hits have probabilities 1/6, 1/20, 1/120, and 1/24. What is the probability of hitting for the cycle if he gets to bat (a) four times, (b) five times? (c) Inequality (2.5) (or (6.4)) from Chapter 1 shows that the answer to (b) is at most 5 times the answer to (a). What is the ratio of the two answers?

1.21. In the World Series, two teams play until one team has won four games. Suppose that the outcome of each game is determined by flipping a coin. What is the probability that the World Series will last (a) seven games, (b) six games, (c) five games, (d) four games?

1.22. Suppose that in the World Series, team B has probability 0.6 of winning each game. Assuming that the outcomes of the games are independent, what is the probability B will win the series?

1.23. David claims to be able to distinguish brand B beer from brand H but Alice claims that he just guesses. They set up a taste test with 10 small glasses of beer. David wins if he gets 8 or more right. What is the probability he will win (a) if he is just guessing? (b) if he gets the right answer with probability 0.9?

1.24. Samuel Pepys wrote to Isaac Newton: "What is more likely, (a) one 6 in 6 rolls of one die or (b) two 6's in 12 rolls?" Compute the probabilities of these events.

1.25. How many children must a couple have before the probability of two boys is larger than 0.75?

1.26. How many times should you roll a die so that the probability of at least one 6 is at least 0.9?

1.27. Suppose we draw k cards out of a deck. What is the probability that we do not draw an Ace? Is the answer larger or smaller than $(3/4)^k$?

2.2. Conditional Probability

Suppose we are told that the event A with $P(A) > 0$ occurs. Then the sample space is reduced from Ω to A and the probability that B will occur given that A has occurred is

(2.1)
$$P(B|A) = P(B \cap A)/P(A)$$

To explain this formula, note that (i) only the part of B that lies in A can possibly occur, and (ii) since the sample space is now A, we have to divide by $P(A)$ to make $P(A|A) = 1$. Some examples should help to clarify the definition.

Example 2.1. Suppose A and B are independent. In this case $P(A \cap B) = P(A)P(B)$ so

$$P(B|A) = \frac{P(A)P(B)}{P(A)} = P(B)$$

Using the words of the intuitive definition of independence, "the occurrence of A has no influence on the probability of the occurrence of B."

Example 2.2. Suppose we roll two dice and $A =$ "The sum is 8," $B =$ "The first die is 3." $A = \{(2,6), (3,5), (4,4), (5,3), (6,2)\}$, so $P(A) = 5/36$.

$A \cap B = \{(3,5)\}$, so

$$P(B|A) = \frac{1/36}{5/36} = \frac{1}{5}$$

The same result holds if B = "The first die is k" and $2 \leq k \leq 6$. Carrying this reasoning further, we see that given the outcome lies in A, all five possibilities have the same probability. This should not be surprising. The original probability is uniform over the 36 possibilities, so when we condition on the occurrence of A, its five outcomes are equally likely.

As the last example may have suggested, the mapping $B \rightarrow P(B|A)$ is a probability. That is, it is a way of assigning numbers to events that satisfies the axioms introduced in Section 1.2 of Chapter 1. To prove this, we note that

(i) $0 \leq P(B|A) \leq 1$ since $0 \leq P(B \cap A) \leq P(A)$.

(ii) $P(\Omega|A) = P(\Omega \cap A)/P(A) = 1$

(iii) If B_i are disjoint then $B_i \cap A$ are disjoint and $(\cup_i B_i) \cap A = \cup_i (B_i \cap A)$, so using the definition of conditional probability and part (iii) of the definition of probability we have

$$P(\cup_i B_i|A) = \frac{P(\cup_i(B_i \cap A))}{P(A)} = \frac{\sum_i P(B_i \cap A)}{P(A)} = \sum_i P(B_i|A)$$

From the last observation it follows that $P(\cdot|A)$ has the same properties that ordinary probabilities do, for example

(2.2) $P(B^c|A) = 1 - P(B|A)$

EXERCISE 2.1. Prove the following generalization of (2.2) directly from the definition: If $(B \cap C) \cap A = \emptyset$ and $(B \cup C) \supset A$ then $P(B|A) + P(C|A) = 1$.

The hypotheses say that from A's point of view $C = B^c$. We will need this result in Example 2.6, which concerns drawing cards out of a deck. There A = "We get at least one Ace," B = "We get exactly one Ace," and C = "We get at least two Aces."

Multiplying the definition of conditional probability in (2.1) on each side by $P(A)$ gives the **multiplication rule**

(2.3) $P(A)P(B|A) = P(B \cap A)$

Example 2.3. Suppose we draw two cards out of a deck of 52. Let A = "The first card is a spade," B = "The second card is a spade." $P(A) = 1/13$. To

$$= \frac{13}{52}$$

compute $P(B|A)$ we note that if A has occurred then only 12 of the remaining 51 cards are spades, so $P(B|A) = 12/51$ and

$$P(A \cap B) = P(A)P(B|A) = \frac{13}{52} \cdot \frac{12}{51}$$

in agreement with the computation in Example 1.4.

Note that in this example we computed $P(B|A)$ by thinking about the situation that exists after A has occurred, rather than using the definition $P(B|A) = P(A \cap B)/P(A)$. Indeed, it is more common to use $P(A)$ and $P(B|A)$ to compute $P(A \cap B)$ than to use $P(A)$ and $P(A \cap B)$ to compute $P(B|A)$.

Example 2.4. In the game of bridge, each of the four players gets 13 cards. If North and South have 8 spades between them, what is the probability that East has 3 spades and West has 2?

We can imagine that first North and South take their 26 cards and then East draws his 13 cards from the 26 that remain. Since there are 5 spades and 21 nonspades, the probability he receives 3 spades and 10 nonspades is $C_{5,3}C_{21,10}/C_{26,13}$. To compute the last probability it is useful to observe that

$$\frac{C_{5,3}\,C_{21,10}}{C_{26,13}} = \frac{\frac{5!}{3!\,2!} \cdot \frac{21!}{10!\,11!}}{\frac{26!}{13!\,13!}} = \frac{\frac{13!}{3!\,10!} \cdot \frac{13!}{2!\,11!}}{\frac{26!}{5!\,21!}} = \frac{C_{13,3}\,C_{13,2}}{C_{26,5}}$$

To arrive at the answer on the right-hand side directly, think of 26 blanks, the first thirteen being East's cards, the second thirteen being West's. We have to pick 5 blanks for spades, which can be done in $C_{26,5}$ ways, while the number of ways of giving East 3 spades and West 2 spades is $C_{13,3}C_{13,2}$. After a lot of cancellation the right-hand side is

$$\frac{13 \cdot 2 \cdot 11 \cdot 13 \cdot 6}{26 \cdot 5 \cdot 2 \cdot 23 \cdot 22} = \frac{22,308}{65,708} = 0.3395$$

Multiplying the last answer by 2, we see that with probability 0.6782 the five outstanding spades will be divided 3–2, that is, one opponent will have 3 and the other 2. Similar computations show that the probabilities of 4–1 and 5–0 splits are 0.2827 and 0.0391.

EXERCISE 2.2. Suppose now that North and South have 9 spades which include the A and K but not the Q. Which event is more likely, $A = $ "East has the Q" or $B = $ "The 4 outstanding spades are split 2–2?" Bridge players will recognize this question as: Should you finesse or play the A and K and hope to capture the Q?

EXERCISE 2.3. In the last problem we ignored the possibility that East has three spades and West has only the Q, since the correct play is to first lead the A and then finesse if the Q does not drop. What is the probability that East has three spades and West has only the Q?

Conditional probabilities are the sources of many "paradoxes" in probability. One that has received some publicity is

Example 2.5. The Monty Hall problem. The problem is named for the host of the television show *Let's Make A Deal* in which contestants were often placed in situations like the following: Three curtains are numbered 1, 2, and 3. Behind one curtain is a car; behind the other two curtains are donkeys. You pick a curtain, say #1. To build some suspense the host opens up one of the two remaining curtains, say #3, to reveal a donkey. What is the probability you will win given that there is a donkey behind #3? Should you switch curtains and pick #2 if you are given the chance?

Many people argue that "the two unopened curtains are the same so they each will contain the car with probability 1/2, and hence there is no point in switching." This naive reasoning is incorrect, but to compute the answer we have to make an assumption about how the host behaves. Suppose that he always chooses to show you a donkey and picks at random if there are two unchosen curtains with donkeys. Assuming you pick curtain #1, there are three possibilities

	#1	#2	#3	host's action
case 1	donkey	donkey	car	opens #2
case 2	donkey	car	donkey	opens #3
case 3	car	donkey	donkey	opens #2 or #3

Now $P(\text{case 2}, \text{open door } \#3) = 1/3$ and

$$P(\text{case 3}, \text{open door } \#3) = P(\text{case 3})P(\text{open door } \#3|\text{case 3}) = \frac{1}{3} \cdot \frac{1}{2} = \frac{1}{6}$$

Adding the two ways door #3 can be opened gives $P(\text{open door } \#3) = 1/2$ and it follows that

$$P(\text{case 3}|\text{open door } \#3) = \frac{P(\text{case 3}, \text{open door } \#3)}{P(\text{open door } \#3)} = \frac{1/6}{1/2} = \frac{1}{3}$$

Although it took a number of steps to compute this answer, it is "obvious." When we picked one of the three doors initially we had probability 1/3 of picking the car, and since the host can always open a door with a donkey the new information does not change our chance of winning.

To further confuse the issue, we would like to observe that we can make the answer to this question 1/2 or 1 depending on what we assume about the host's behavior. If we assume that the host does not know what is behind the curtains and opens one that you did not choose at random then the answer is 1/2, since in this case the two unopened curtains play symmetric roles. If we assume that the host acts to minimize the cost of the show by opening your curtain immediately when you are wrong, then the answer is 1 since he will only show you a donkey if you have picked the car.

EXERCISE 2.4. Three prisoners, Al, Bob, and Charlie, are in a cell. At dawn two will be set free and one will be hanged, but they do not know who will be chosen. The guard offers to tell Al the name of one of the other two prisoners who will go free but Al stops him, screaming, "No, don't! That would increase my chances of being hanged to 1/2." Criticize Al's reasoning.

Example 2.6. A person picks 13 cards out of a deck of 52. Let $A_1 =$ "He receives at least one Ace," $B =$ "He has the Ace of hearts," and $A_2 =$ "He receives at least two Aces." Since all Aces are alike, it may at first be surprising that the probability he has two Aces given that he has the Ace of hearts is larger than the probability he has two Aces given that he has one Ace, but this is true.

Let $E_1 =$ "He has exactly one Ace." Exercise 2.1 implies that

$$P(A_2|A_1) = 1 - P(E_1|A_1) \qquad P(A_2|B) = 1 - P(E_1|B)$$

so it suffices to show that $P(E_1|A_1) > P(E_1|B)$. Let $E_0 =$ "He has no Ace."

$$p_0 = P(E_0) = C_{48,13}/C_{52,13} \qquad p_1 = P(E_1) = 4C_{48,12}/C_{52,13}$$

Since $E_1 \subset A_1$, $P(E_1|A_1) = P(E_1)/P(A_1) = p_1/(1 - p_0)$. On the other hand,

$$P(E_1|B) = \frac{P(E_1 \cap B)}{P(B)} = \frac{C_{48,12}/C_{52,13}}{1/4} = p_1$$

so $P(E_1|A_1) = p_1/(1 - p_0) > p_1 = P(E_1|B)$ as claimed. The intuitive explanation of the last result is that it is harder to get the Ace of hearts than to get at least one Ace, so conditioning on having the Ace of hearts gives you a better chance of having two or more Aces than conditioning on having at least one Ace.

EXERCISES

2.5. A friend flips two coins and tells you that at least one is Heads. Given this information, what is the probability that the first coin is Heads?

2.6. A friend rolls two dice and tells you that there is at least one 6. What is the probability the sum is at least 9?

2.7. Suppose you draw five cards out of a deck of 52 and get 2 spades and 3 hearts. What is the probability the first card drawn was a spade?

2.8. We draw 4 cards out of a deck of 52. Find the probability that all four values are different given that (a) the four suits are different, (b) we drew two spades and two hearts.

2.9. Two people, whom we call South and North, draw 13 cards out of a deck of 52. South has two Aces. What is the probability that North has (a) none? (b) one? (c) the other two?

2.10. An urn contains 8 red, 7 blue, and 5 green balls. You draw out two balls and they are different colors. Given this, what is the probability the two balls were red and blue?

2.11. Suppose 60% of the people subscribe to newspaper A, 40% to newspaper B, and 30% to both. If we pick a person at random who subscribes to at least one newspaper, what is the probability she subscribes to newspaper A?

2.12. 80% of people who start school graduate from high school, 60% of high school graduates graduate from college, and 15% of college graduates get an advanced degree. What percent of people who start school get an advanced degree?

2.13. Suppose that the probability a married man votes is 0.45, the probability a married woman votes is 0.4, and the probability a woman votes given that her husband does is 0.6. What is the probability (a) both vote, (b) a man votes given that his wife does?

2.14. Two events have $P(A) = 1/4$, $P(B|A) = 1/2$, and $P(A|B) = 1/3$. Compute $P(A \cap B)$, $P(B)$, $P(A \cup B)$.

2.15. A, B, and C are events with $P(A) = 0.3$, $P(B) = 0.4$, $P(C) = 0.5$, A and B are disjoint, A and C are independent, and $P(B|C) = 0.1$. Find $P(A \cup B \cup C)$.

2.16. Show that $P(A \cap B \cap C) = P(A)P(B|A)P(C|A \cap B)$.

2.17. Show that $P(E|A) = P(E|A \cap B)P(B|A) + P(E|A \cap B^c)P(B^c|A)$.

2.3. Two-Stage Experiments

We begin with an example and then describe the collection of problems we will treat in this section.

Example 3.1. Al flips 3 coins and Betty flips 2. Al wins if the number of Heads he gets is more than the number Betty gets. What is the probability Al will win?

Let A be the event that Al wins, let B_i be the event that Betty gets i Heads, and let C_j be the event that Al gets j Heads. By considering the four outcomes of flipping two coins it is easy to see that

$$P(B_0) = 1/4 \qquad P(B_1) = 1/2 \qquad P(B_2) = 1/4$$

while considering the eight outcomes for three coins leads to

$$P(A|B_0) = P(C_1 \cup C_2 \cup C_3) = 7/8$$
$$P(A|B_1) = P(C_2 \cup C_3) = 4/8$$
$$P(A|B_2) = P(C_3) = 1/8$$

Since $A \cap B_i$, $i = 0, 1, 2$ are disjoint and their union is A, we have

$$P(A) = \sum_{i=0}^{2} P(A \cap B_i) = \sum_{i=0}^{2} P(A|B_i)P(B_i)$$

since $P(A \cap B_i) = P(A|B_i)P(B_i)$ by the multiplication rule (2.3). Plugging in the values we computed,

$$P(A) = \frac{1}{4} \cdot \frac{7}{8} + \frac{2}{4} \cdot \frac{4}{8} + \frac{1}{4} \cdot \frac{1}{8} = \frac{7 + 8 + 1}{32} = \frac{1}{2}$$

EXERCISE 3.1. The last calculation makes it look miraculous that we have a fair game. The reason becomes clear when we look at a more general problem. Suppose Al flips $n + 1$ coins and Betty flips n coins. By considering the three possibilities $A > B$, $A = B$, $A < B$ that can occur after Al has flipped n coins and Betty has flipped n, show that the probability Al has more ($>$) Heads than Betty is $1/2$.

Abstracting the structure of the last problem, let $B_1, \ldots, B_k$ be a **partition**, that is, a collection of disjoint sets whose union is Ω. Using the fact that the sets $A \cap B_i$ are disjoint, and the multiplication rule, we have

$$(3.1) \qquad P(A) = \sum_{i=1}^{k} P(A \cap B_i) = \sum_{i=1}^{k} P(A|B_i)P(B_i)$$

a formula that is sometimes called the **law of total probability**.

The name of this section comes from the fact that we think of our experiment as occurring in two stages. The first stage determines which of the B's occur, and when B_i occurs in the first stage A occurs with probability $P(A|B_i)$ in the second. As the next example shows, the two stages are sometimes clearly visible in the problem itself.

Example 3.2. Roll a die and then flip that number of coins. What is the probability of A = "We get exactly 3 Heads"?

Let B_i = "The die shows i." $P(B_i) = 1/6$ for $i = 1, 2, \ldots, 6$ and

$$P(A|B_1) = 0 \qquad P(A|B_2) = 0 \qquad P(A|B_3) = 2^{-3}$$
$$P(A|B_4) = C_{4,3}\, 2^{-4} \qquad P(A|B_5) = C_{5,3}\, 2^{-5} \qquad P(A|B_6) = C_{6,3}\, 2^{-6}$$

So plugging into (3.1),

$$P(A) = \frac{1}{6}\left\{ \frac{1}{8} + \frac{4}{16} + \frac{10}{32} + \frac{20}{64} \right\}$$
$$= \frac{1}{6}\left\{ \frac{8 + 16 + 20 + 20}{64} \right\} = \frac{1}{6}$$

In the same way we can compute the probability of A_k = "We get exactly k Heads":

k	0	1	2	3	4	5	6
$P(A_k)$	$\frac{63}{384}$	$\frac{120}{384}$	$\frac{99}{384}$	$\frac{64}{384}$	$\frac{29}{384}$	$\frac{8}{384}$	$\frac{1}{384}$

EXERCISE 3.2. Verify the entry for $k = 1$.

Example 3.3. Suppose we roll three dice. What is the probability that the sum is 9?

Let A = "The sum is 9," B_i = "The first die shows i," and C_j = "The sum of the second and third dice is j." Now $P(A|B_i) = P(C_{9-i})$ and we know the probabilities for the sum of two dice:

j	2	3	4	5	6	7	8	9	10	11	12
$P(C_j)$	$\frac{1}{36}$	$\frac{2}{36}$	$\frac{3}{36}$	$\frac{4}{36}$	$\frac{5}{36}$	$\frac{6}{36}$	$\frac{5}{36}$	$\frac{4}{36}$	$\frac{3}{36}$	$\frac{2}{36}$	$\frac{1}{36}$

Using (3.1), now we have

$$P(A) = \sum_{i=1}^{6} P(B_i)P(A|B_i) = \frac{1}{6}\left(P(C_8) + P(C_7) + \cdots + P(C_3) \right)$$
$$= \frac{1}{6}\left(\frac{5}{36} + \frac{6}{36} + \frac{5}{36} + \frac{4}{36} + \frac{3}{36} + \frac{2}{36} \right) = \frac{25}{216}$$

In the same way we can compute the probability of A_k = "The sum of three dice is k":

k	3,18	4,17	5,16	6,15	7,14	8,13	9,12	10,11
$P(A_k)$	$\frac{1}{216}$	$\frac{3}{216}$	$\frac{6}{216}$	$\frac{10}{216}$	$\frac{15}{216}$	$\frac{21}{216}$	$\frac{25}{216}$	$\frac{27}{216}$

EXERCISE 3.3. Verify the entry for $k = 6$.

With patience one can use the idea in this example to compute $p_m(k) =$ the probability that the sum of m dice is k. Breaking things down according to the number on the first die and using (3.1), we have

$$p_m(k) = \sum_{j=k-6}^{k-1} \frac{p_{m-1}(j)}{6}$$

EXERCISE 3.4. Compute $p_4(12)$ by using the last formula or by breaking things down according to the sum of the first two dice.

Example 3.4. Craps. As an application of (3.1) we will now compute the probability that the "shooter" (the person rolling the two dice used) wins in the game of craps. In this game, if the sum of the dice is 2, 3, or 12 on his first roll, he loses; if the sum is 7 or 11, he wins; if the sum is 4, 5, 6, 8, 9, or 10, this number becomes his "point" and he wins if he "makes his point," i.e., his number comes up again before he throws a 7. The first step in analyzing craps is to compute the probability that the shooter makes his point. Suppose his point is 5 and let E_k be the event that the sum is k. There are 4 outcomes in E_5 $((1,4), (2,3), (3,2), (4,1))$, 6 in E_7, and hence 26 not in $E_5 \cup E_7$. Letting $\times$ stand for "The sum is not 5 or 7," we see that

$$P(5) = \frac{4}{36} \quad P(\times\, 5) = \frac{26}{36} \cdot \frac{4}{36} \quad P(\times \times 5) = \left(\frac{26}{36}\right)^2 \frac{4}{36}$$

From the first three terms it is easy to see that for $k \geq 0$

$$P(\times \text{ on } k \text{ rolls then } 5) = \left(\frac{26}{36}\right)^k \frac{4}{36}$$

Summing over the possibilities, which represent disjoint ways of rolling 5 before 7, we have

$$P(5 \text{ before } 7) = \sum_{k=0}^{\infty} \left(\frac{26}{36}\right)^k \frac{4}{36} = \frac{4}{36} \cdot \frac{1}{1 - \frac{26}{36}}$$

since

(3.2)
$$\sum_{k=0}^{\infty} x^k = \frac{1}{1-x}$$

Simplifying, we have $P(5 \text{ before } 7) = (4/36)/(10/36) = 4/10$. Such a simple answer should have a simple explanation, and it does. If we ignore the outcomes that result in a sum other than 5 or 7, we reduce the sample space from Ω to $E = E_5 \cup E_7$ and the distribution of the first outcome that lands in E follows the conditional probability $P(\cdot|E)$. Since $E_5 \cap E = E_5$ we have

$$P(E_5|E) = \frac{P(E_5)}{P(E)} = \frac{4/36}{10/36} = \frac{4}{10}$$

The last argument generalizes easily to give the probabilities of making any point

k	4	5	6	8	9	10
$|E_k|$	3	4	5	5	4	3
P(k before 7)	3/9	4/10	5/11	5/11	4/10	3/9

To compute the probability of $A =$ "He wins," we let $B_k =$ "The first roll is k," and observe that (3.1) implies

$$P(A) = \sum_{k=2}^{12} P(A \cap B_k) = \sum_{k=2}^{12} P(B_k)P(A|B_k)$$

When $k = 2, 3,$ or 12 comes up on the first roll we lose, so

$$P(A|B_k) = 0 \quad \text{and} \quad P(A \cap B_k) = 0$$

When $k = 7$ or 11 comes up on the first roll we win, so

$$P(A|B_k) = 1 \quad \text{and} \quad P(A \cap B_k) = P(B_k)$$

When the first roll is $k = 4, 5, 6, 8, 9,$ or 10, $P(A|B_k) = P(k \text{ before } 7)$ and $P(A \cap B_k)$ is

$$\frac{3}{36} \cdot \frac{3}{9} \quad k = 4, 10 \qquad \frac{4}{36} \cdot \frac{4}{10} \quad k = 5, 9 \qquad \frac{5}{36} \cdot \frac{5}{11} \quad k = 6, 8$$

Adding up the terms in the sum in the order in which they were computed,

$$P(A) = \frac{6}{36} + \frac{2}{36} + 2\left(\frac{1}{36} + \frac{4 \cdot 2}{36 \cdot 5} + \frac{5 \cdot 5}{36 \cdot 11}\right)$$

$$= \frac{4}{18} + 2\left(\frac{55 + 88 + 125}{36 \cdot 11 \cdot 5}\right) = \frac{220 + 268}{18 \cdot 11 \cdot 5} = \frac{488}{990} = 0.4929$$

which is not very much less than $1/2 = 495/990$.

Example 3.5. Genetics. Most animals and plants are diploid organisms: each cell has two copies of each chromosome, with the exception of the chromosome that determines the individual's sex. In this case, a female has two copies of the X chromosone and a male has one X and one Y. When reproduction occurs, a special cell division process called *meiosis* produces reproductive cells called *gametes* that have one copy of each chromosome. Two gametes are then combined to produce one new individual.

Each hereditary characteristic is carried by a pair of genes, one on each chromosome, so the new offspring gets one gene from its mother and one from its father. We will consider the situation in which each gene can take only two forms, called *alleles*, which we will denote by a and A. An example from the pioneering work of the Czech monk Gregor Mendel is $A =$ "smooth skin" and $a =$ "wrinkled skin" for the pea plants that he used for much of his experimental work. In this case A is *dominant* over a, meaning that Aa individuals (those with one A and one a) will have smooth skin.

Let us start from an idealized infinite population in which individuals are found in the following proportions, where the proportions are nonnegative and sum to 1:

$$\begin{array}{ccc} AA & Aa & aa \\ \alpha_0 & \beta_0 & \gamma_0 \end{array}$$

If we assume that random mating occurs then each new individual picks two parents at random from the population and picks an allele at random from the two carried by each parent. To compute the proportions of the three types in the first generation of offspring, note that (i) since the first allele is picked at random from the population it will be A with probability

$$p_1 = \alpha_0 + (\beta_0/2)$$

and a with probability $1 - p_1 = \gamma_0 + (\beta_0/2)$, and (ii) the second allele will be independent and have the same distribution, so the proportions in the first generation of offspring will be

$$\alpha_1 = p_1^2 \qquad \beta_1 = 2p_1(1 - p_1) \qquad \gamma_1 = (1 - p_1)^2$$

Something quite remarkable happens when we use these values to compute the fractions in the second generation of offspring. An allele picked at random from the first generation will be A with probability

$$p_2 = \alpha_1 + (\beta_1)/2$$
$$= p_1^2 + 2p_1(1 - p_1)/2 = p_1(p_1 + 1 - p_1) = p_1$$

so the proportions in the second generation of offspring will be

$$\alpha_2 = p_2^2 = p_1^2 = \alpha_1$$
$$\beta_2 = 2p_2(1 - p_2) = 2p_1(1 - p_1) = \beta_1$$
$$\gamma_2 = (1 - p_2)^2 = (1 - p_1)^2 = \gamma_1$$

Since the proportions of AA, Aa, and aa alleles reach equilibrium in one generation of offspring starting from an arbitrary distribution, it follows that if the fraction of A alleles in the population is p then the proportions of the genotypes will be

AA	Aa	aa
p^2	$2p(1 - p)$	$(1 - p)^2$

The last result is called the *Hardy-Weinberg Theorem*. To illustrate its use suppose that in a population of pea plants, 91% have smooth skin (AA or Aa) and 9% have wrinkled skin (aa). Since the fractions of AA, Aa, and aa individuals are p^2, $2p(1-p)$, and $(1-p)^2$ and only aa individuals have wrinkled skin, we can infer that $(1 - p) = 0.3$ and the three proportions must be 0.49, 0.42, and 0.09.

EXERCISES

3.5. An urn has 12 red and 8 black balls. (a) By considering the possibilities for the first draw, compute the probability the second ball drawn is red. (b) Repeat part (a) for the third draw. (c) What is the probability the 10th ball drawn is red?

3.6. The population of Cyprus is 70% Greek and 30% Turkish. 20% of the Greeks and 10% of the Turks speak English. What fraction of the people of Cyprus speak English?

3.7. You are going to meet a friend at the airport. Your experience tells you that the plane is late 70% of the time when it rains, but is late only 20% of the time when it does not rain. The weather forecast that morning calls for a 40% chance of rain. What is the probability the plane will be late?

3.8. Two boys have identical piggy banks. The older boy has 18 quarters and 12 dimes in his; the younger boy, 2 quarters and 8 dimes. One day the two banks get mixed up. You pick up a bank at random and shake it until a coin comes out. What is the probability you get a quarter? Note that there are 20 quarters and 20 dimes in all.

3.9. How can 5 black and 5 white balls be put into two urns to maximize the probability a white ball is drawn when we draw from a randomly chosen urn?

3.10. A student is taking a multiple-choice test in which each question has four possible answers. She knows the answers to 50% of the questions, can narrow the choices down to two 30% of the time, and does not know anything about 20% of the questions. What is the probability she will correctly answer a question chosen at random from the test?

3.11. A student is taking a multiple-choice test in which each question has four possible answers. She knows the answers to 5 of the questions, can narrow the choices down in 3 cases, and does not know anything about 2 of the questions. What is the probability she will correctly answer (a) 10, (b) 9, (c) 8, (d) 7, (e) 6, (f) 5 questions?

3.12. From a signpost that says MIAMI two letters fall off. A friendly drunk puts the two letters back into the two empty slots at random. What is the probability that the sign still says MIAMI?

3.13. A group of n people is going to sit in a row of n chairs. By considering the chair Mr. Jones sits in, compute the probability Mr. Jones will sit next to Miss Smith.

3.14. Suppose we roll two dice twice. What is the probability we get the same sum on both throws?

3.15. Give yet another proof of $P(5$ before $7) = 4/10$ by arguing that if $p = P(5$ before $7)$ then $p = 4/36 + (26/36)p$.

3.16. *Huygens' problem.* Two boys, Charlie and Doug, take turns rolling two dice with Charlie going first. If Charlie rolls a 6 before Doug rolls a 7 he wins. What is the probability Charlie wins?

3.17. Two boys take turns throwing darts at a target. Al throws first and hits with probability 1/4. Bob throws second and hits with probability 1/3. What is the probability Al will hit the target before Bob does?

3.18. Three boys take turns shooting a basketball and have probabilities 0.2, 0.3, and 0.5 of scoring a basket. Who has the best chance of getting the first basket?

3.19. Change the second and third probabilities in the last problem so that each boy has an equal chance of winning.

Example 3.6. Two-state Markov chains. Suppose that there are two brands of toothpaste, A and B, and that on each purchase a customer changes brands with probability 1/4 and buys the same brand with probability 3/4. Let f_k be the fraction that buy brand A on their kth purchase and suppose $f_1 = 1/3$. Compute $f_2, f_3, \ldots$

Now to buy brand A on the kth purchase, the customer must either buy brand A at time $k - 1$ and stay loyal, an event of probability $(3/4)f_{k-1}$, or buy brand B at time $k - 1$ and switch, an event of probability $(1/4)(1 - f_{k-1})$. Combining the last two observations, we have

$$(\star) \qquad f_k = \frac{3}{4}f_{k-1} + \frac{1}{4}(1 - f_{k-1})$$

Using this formula repeatedly, one finds

k	1	2	3	4	5
f_k	$\frac{2}{6}$	$\frac{5}{12}$	$\frac{11}{24}$	$\frac{23}{48}$	$\frac{47}{96}$
$1/2 - f_k$	$\frac{1}{6}$	$\frac{1}{12}$	$\frac{1}{24}$	$\frac{1}{48}$	$\frac{1}{96}$

which suggests that the f_k are converging to $1/2$ and furthermore that $f_k = 1/2 - 1/(3 \cdot 2^k)$. To prove this we subtract $1/2$ from each side of $(\star)$ to get

$$f_k - 1/2 = \frac{3}{4}(f_{k-1} - 1/2) + \frac{1}{4}(1/2 - f_{k-1}) = \frac{1}{2}(f_{k-1} - 1/2)$$

In words, the distance from f_k to $1/2$ decreases by a factor of 2 each time.

Generalizing from the last example, we can consider a system that has two states (here "buy brand A" or "buy brand B") and has the property that if we are in state A at time $k - 1$ then we will stay in state A with probability $1 - p$ and switch to state B with probability p, whereas if we are in state B we will stay in state B with probability $1 - q$ and switch to state A with probability q. The phrase "Markov chain" refers to the fact that we assume that the state at time k depends only on the state at time $k - 1$ and not on the states at earlier times.

To avoid trivialities we will suppose that switching and staying put are both possible, or, to be precise, $0 < p + q < 2$. If we let f_k be the probability of being in state A at time k then we have

$$(\star\star) \qquad f_k = (1 - p)f_{k-1} + q(1 - f_{k-1})$$

To investigate the limiting behavior of f_k we will first look for a probability π so that if $f_{k-1} = \pi$ then $f_k = \pi$ (and then $f_{k+1} = \pi$ and ...). To do this we set

$$\pi = (1 - p)\pi + q(1 - \pi)$$

and solve to get $\pi = q/(p + q)$. Subtracting the last two equations we have

$$f_k - \pi = (1 - p)(f_{k-1} - \pi) + q(\pi - f_{k-1}) = (1 - p - q)(f_{k-1} - \pi)$$

Our assumption that $0 < p + q < 2$ implies that $|p + q - 1| < 1$, so $|f_k - \pi|$ decreases to 0 exponentially fast.

3.20. A young woman wants to make her date wait occasionally but does not want him to get too angry. Therefore, if she made him wait the previous time, she is only late with probability 0.2, but if she was not late the previous time, she is late with probability 0.6. Let f_k be the probability she is late on the kth date and suppose that $f_1 = 1$. Compute f_2, f_3, and $\lim_{k \to \infty} f_k$.

3.21. Despite the widely held belief in the "hot hand," a statistical study of the Philadelphia 76ers revealed that they were less likely to make a shot right after they had made the previous one. For example, Darryl Dawkins made 71% of his shots after he had missed once but only 57% after he had made the previous shot. Assuming that the probability of making a shot just depends upon the outcome of the previous shot, compute the long-run fraction of shots Dawkins will make.

3.22. Suppose that a rainy day is followed by another rainy day with probability $1/2$, but a sunny day is followed by a sunny day with probability $2/3$. What is the long-run fraction of sunny days in this simple-minded model of the weather?

2.4. Bayes Formula

The title of the section is a little misleading since we will regard Bayes formula as a method for computing conditional probabilities and will only reluctantly give the formula after we have done four examples to illustrate the method.

Example 4.1. Exit polls. In the California gubernatorial election in 1982, several TV stations predicted, on the basis of questioning people when they exited the polling place, that Tom Bradley, then mayor of Los Angeles, would win the election. When the votes were counted, however, he lost by a considerable margin. What happened?

To give our explanation we need some notation and some numbers. Suppose we choose a person at random, let $B =$ "The person votes for Bradley" and suppose that $P(B) = 0.45$. There were only two candidates, so this makes the probability of voting for Deukmejian $P(B^c) = 0.55$. Let $A =$ "The voter stops and answers a question about how she voted" and suppose that $P(A|B) = 0.4$, $P(A|B^c) = 0.3$. That is, 40% of Bradley voters will respond compared to 30% of the Deukmejian voters. We are interested in computing $P(B|A) =$ the fraction of voters in our sample that voted for Bradley. By the definition of conditional probability (2.1),

$$P(B|A) = P(B \cap A)/P(A)$$

To evaluate the numerator we use the multiplication rule (2.3)

$$P(B \cap A) = P(B)P(A|B) = 0.45 \cdot 0.4 = 0.18$$

Similarly,
$$P(B^c \cap A) = P(B^c)P(A|B^c) = 0.55 \cdot 0.3 = 0.165$$
Now $P(A) = P(B \cap A) + P(B^c \cap A)$ so
$$P(B|A) = \frac{P(B \cap A)}{P(A)} = \frac{0.18}{0.18 + 0.165} = 0.5217$$

and from our sample it looks as if Bradley will win. The problem with the exit poll is that the difference in the response rates makes our sample not representative of the population as a whole. Turning to the mechanics of the computation, note that 18% of the voters are for Bradley and respond, while 16.5% are for Deukmejian and respond, so the fraction of Bradley voters in our sample is $18/(18 + 16.5)$. In symbols,

$$P(B|A) = \frac{P(A \cap B)}{P(A \cap B) + P(A \cap B^c)}$$

In words, there are two ways an outcome can be in A – it can be in B or in B^c – and the conditional probability is the fraction of the total that comes from the first way.

Example 4.2. The alpha fetal protein test is meant to detect spina bifida in unborn babies, a condition that affects 1 out of 1000 children who are born. Let B be the event that the baby has spina bifida and B^c be the event that it does not. The literature on the test indicates that 5% of the time a healthy baby will cause a positive reaction. We will assume that the test is positive 100% of the time when spina bifida is present. Your doctor has just told you that your alpha fetal protein test was positive. What is the probability that your baby has spina bifida?

Let $A =$ "a positive reaction." We want to calculate $P(B|A)$. By the definition of conditional probability (2.1),

$$P(B|A) = P(B \cap A)/P(A)$$

To evaluate the numerator we use the multiplication rule (2.3)

$$P(B \cap A) = P(B)P(A|B) = 0.001 \cdot 1 = 0.001$$

Similarly,
$$P(B^c \cap A) = P(B^c)P(A|B^c) = 0.999 \cdot 0.05 \approx 0.05$$
Now $P(A) = P(B \cap A) + P(B^c \cap A)$ so

$$P(B|A) = \frac{P(B \cap A)}{P(A)} = \frac{0.001}{0.001 + 0.050} = \frac{1}{51}$$

Thus the probability of spina bifida given the positive reaction is only about 2%. This situation comes about because it is much easier to have a positive reaction by having a healthy baby and then having a positive reaction, which has probability 0.05, than by having a baby with spina bifida, which has probability 0.001.

We would like to point out that while the conditional probability of spina bifida given a positive reaction is only 2%, this does not mean that the test is worthless. To introduce some terminology from Bayesian statistics, the **prior probability** (i.e., before the test) of spina bifida is 0.1%, whereas the **posterior probability** (i.e., after the test results are known) is about 2%. That is, the probability is now 20 times larger. The positive reaction is thus a warning that more accurate (and more expensive) tests should be done to see if the baby has spina bifida.

Example 4.3. A woman has a brother with hemophilia but two parents who do not have the disease. Since hemophilia is caused by a recessive gene h on the X chromosome, we can infer that her mother is a carrier (that is, the mother has the hemophilia gene h on one of her X chromosomes and the healthy gene H on the other), while her father has the healthy gene on his one X chromosome. Since the woman received one X chromosome from her father and one from her mother, there is a 50% chance that she is a carrier, and if so, there is a 50% chance that her sons will have the disease. If she has two sons without the disease, what is the probability she is a carrier?

Let B be the event that she is a carrier and A be the event that she has two healthy sons. We want to compute $P(B|A)$. By the definition of conditional probability (2.1),

$$P(B|A) = P(B \cap A)/P(A)$$

To evaluate the numerator we use the multiplication rule (2.3)

$$P(B \cap A) = P(B)P(A|B) = \frac{1}{2} \cdot \frac{1}{4} = \frac{1}{8}$$

since the probability of having two healthy sons when she is a carrier is 1/4. Similarly,

$$P(B^c \cap A) = P(B^c)P(A|B^c) = \frac{1}{2} \cdot 1 = \frac{1}{2}$$

Now $P(A) = P(B \cap A) + P(B^c \cap A)$ so

$$P(B|A) = \frac{P(B \cap A)}{P(A)} = \frac{1/8}{1/8 + 1/2} = \frac{1}{5}$$

EXERCISE 4.1. What was the conditional probability she was a carrier after her first son was born? What would it be if she had a third healthy son?

Our final example is a situation with more than two events. Our first assumption is not very realistic but it makes the computations simple.

Example 4.4. Suppose for simplicity that the number of children in a family is 1, 2, or 3, with probability 1/3 each. Little Bobby has no brothers. What is the probability he is an only child?

Let B_1, B_2, B_3 be the events that a family has one, two, or three children, and let A be the event that a family has only one boy. We want to compute $P(B_1|A)$. By the definition of conditional probability (2.1),

$$P(B_1|A) = P(B_1 \cap A)/P(A)$$

To evaluate the numerator we use the multiplication rule (2.3)

$$P(B_1 \cap A) = P(B_1)P(A|B_1) = \frac{1}{3} \cdot \frac{1}{2} = \frac{1}{6}$$

Similarly, $P(B_2 \cap A) = P(B_2)P(A|B_2) = \frac{1}{3} \cdot \frac{2}{4} = \frac{1}{6}$ and

$$P(B_3 \cap A) = P(B_3)P(A|B_3) = \frac{1}{3} \cdot \frac{3}{8} = \frac{1}{8}$$

Now $P(A) = \sum_i P(B_i \cap A)$ so

$$P(B_1|A) = \frac{P(B_1 \cap A)}{P(A)} = \frac{1/6}{1/6 + 1/6 + 1/8} = \frac{8}{8 + 8 + 6} = \frac{4}{11}$$

You may be surprised to find that the next problem has a different answer.

EXERCISE 4.2. Suppose for simplicity that number of children in a family is 1, 2, or 3, with probability 1/3 each. Little Bobby has no sisters. What is the probability he is an only child?

We are now ready to generalize from the four examples and state **Bayes formula.** In each case, we have a **partition** of the probability space $B_1, \ldots, B_n$, i.e., a sequence of disjoint sets with $\cup_{i=1}^{n} B_i = \Omega$. (In the first three examples, $B_1 = B$ and $B_2 = B^c$.) We are given $P(B_i)$ and $P(A|B_i)$ for $1 \le i \le n$ and we want to compute $P(B_1|A)$. By the definition of conditional probability (2.1),

$$P(B_1|A) = P(B_1 \cap A)/P(A)$$

To evaluate the numerator and denominator we observe that

$$P(B_i \cap A) = P(B_i)P(A|B_i)$$

and $P(A) = \sum_i P(B_i \cap A)$ so

(4.1) $$P(B_1|A) = \frac{P(B_1 \cap A)}{P(A)} = \frac{P(B_1)P(A|B_1)}{\sum_i P(B_i)P(A|B_i)}$$

This is Bayes formula. Even though we have numbered it, we advise you not to memorize it. It is much better to remember the procedures we followed to compute the conditional probability.

Example 4.5. Paternity probabilities. Blood testing for hereditary factors is being used increasingly in paternity cases to infer, using Bayes formula, that a particular man is likely to be the father. For a concrete example suppose that the baby's blood type is B, the mother's is A, and that of the suspected father, whom for convenience we will call Bob, is B. To explain how this could happen, we note that the genes that control blood type can be O, A, or B, with A and B dominant over O but neither A nor B dominating the other, so we get the following correspondence between genotypes (the genes on the two chromosomes) and phenotypes (observed blood type):

genotype	OO	AO	AA	BO	BB	AB
phenotype	O	A	A	B	B	AB
proportion	.479	.310	.050	.116	.007	.038

From this table, we see that if the baby's blood type is B then it must be the case that the mother's genotype is AO, she contributed an O gene, and the father contributed a B gene.

Let E (for "evidence") be the event that the baby's blood type is B, and F be the event that Bob is indeed the father. We cannot observe Bob's genotype, but using the proportions of the various genotypes from the table (which were inferred from the observed proportion of phenotypes using a generalization of the Hardy-Weinberg law, discussed in Example 3.5) we can compute that

$$P(\text{genotype is } BO | \text{phenotype is } B) = 0.116/0.123$$

$$P(E|F) = \frac{(0.116)0.5 + 0.007}{0.123} = \frac{0.065}{0.123} = 0.528$$

There is not too much to argue about in the last computation. When we compute $P(E|F^c)$, we make the first of two questionable assumptions: If Bob is not the father, then the real father is someone chosen at random from the population, so

$$P(E|F^c) = (0.116)0.5 + 0.007 = 0.065$$

Taking the ratio of the conditional probabilities gives

$$\frac{P(E|F)}{P(E|F^c)} = \frac{1}{0.123} = 8.13$$

a number that is called the **paternity index** and is interpreted in this case as saying that Bob is 8.13 times as likely to be the father as a man chosen at random from the population. The reason for interest in this number comes from the fact that by turning Bayes' formula (4.1) upside down we have

$$(4.2) \qquad P(F|E) = \left(1 + \frac{P(F^c)}{P(F)} \cdot \frac{P(E|F^c)}{P(E|F)}\right)^{-1}$$

To evaluate $P(F|E)$ we thus need to evaluate the **prior probability** $P(F)$ that Bob is the father. It would be natural to make $P(F)$ equal to the fraction of times that the mother had intercourse with Bob near the time of conception. However, it is not unusual for the mother to claim this number is 1 and the alleged father to claim it is 0, so the common practice in these computations is to set $P(F) = 1/2$ (our second questionable assumption). If we do this, it follows that

$$P(F|E) = (1.123)^{-1} = 0.8904$$

In addition to blood type, there are a number of other aspects of the blood that can be tested to get genetic information. Suppose now that new evidence E' regarding the so-called Gc serum protein is introduced. We will not describe this quantity except to say that there are two alleles 1 and 2, which are co-dominant, so there are three phenotypes that have the indicated frequencies:

phenotype	11	12	22
frequency	0.504	0.412	0.084

Suppose now that the child's phenotype is 12, the mother's is 11, and Bob's is 12. For the child's to be 12, the father must have contributed the 2, so

$$P(E'|F) = 0.5$$

while if the father is chosen at random,

$$P(E'|F^c) = 0.412(0.5) + 0.084(1) = 0.290$$

To combine this with the earlier information, we will now make the assumption that the choice of blood group allele and the choice of the Gc serum protein allele that the father gives to the child are made independently. This is clear if the genes for blood group and Gc serum protein are on different chromosomes. However, it is also a reasonable assumption if the genes lie on the same chromosome, since the B blood group allele has an equal chance to be paired with the 1 or the 2 Gc allele. From the last assumption, it follows that

$$P(E'|E \cap F) = P(E'|F) \qquad P(E'|E \cap F^c) = P(E'|F^c)$$

So using the definition of conditional probability, the formula in Exercise 2.16, and our last observation,

$$\frac{P(F|E \cap E')}{P(F^c|E \cap E')} = \frac{P(F \cap E \cap E')}{P(F^c \cap E \cap E')} = \frac{P(F)}{P(F^c)} \cdot \frac{P(E|F)}{P(E|F^c)} \cdot \frac{P(E'|F \cap E)}{P(E'|F^c \cap E)}$$

$$= \frac{P(F)}{P(F^c)} \cdot \frac{P(E|F)}{P(E|F^c)} \cdot \frac{P(E'|F)}{P(E'|F^c)}$$

In words, the new paternity index is obtained from the old one by multiplying by $P(E'|F)/P(E'|F^c) = 0.5/0.29 = 1.724$. This raises the paternity index to $8.13 \cdot 1.724 = 14.016$, and using a generalization of (4.2) the new conditional probability that Bob is the father is

$$P(F|E \cap E') = \left(1 + \frac{1}{14.016}\right)^{-1} = (1.0713)^{-1} = 0.933$$

More tests could drive Bob further up **Hummel's Likelihood of Paternity** scale–

Plausibility of Paternity	Likelihood of Paternity
0.9980–0.9990	Practically proved
0.9910–0.9979	Extremely likely
0.9500–0.9909	Very likely
0.9000–0.9499	Likely
0.8000–0.8999	Undecided
Less than 0.8000	Not useful

–or of course they could exonerate him. The scale printed here, as well as most of the facts cited above, are from *Statistics and the Law*, by M.H. de Groot, S.B. Fienberg, and J.B. Kadane, published in 1986 by John Wiley and Sons.

EXERCISE 4.3. By using ideas in Example 3.5, show that if p_O, p_A, and p_B are the fraction of O, A, and B genes in the population then the fractions of phenotypes O, A, B, and AB are p_O^2, $2p_B p_O + p_B^2$, $2p_A p_O + p_A^2$, and $p_A p_B$. Use the observed phenotypic frequencies 0.479, 0.360, 0.123, and 0.038 to compute p_O, p_A, and p_B and confirm the genotypic frequencies given in the table above.

EXERCISE 4.4. A and B are said to be **conditionally independent given F** if $P(A \cap B|F) = P(A|F)P(B|F)$. Assuming $P(A \cap F) > 0$, show that this is equivalent to $P(B|A \cap F) = P(B|F)$.

EXERCISES

4.5. 5% of men and 0.25% of women are colorblind. What is the probability a colorblind person is a man?

4.6. Binary digits, i.e., 0's and 1's, are sent down a noisy communications channel. They are received as sent with probability 0.9 but errors occur with probability 0.1. Assuming that 0's and 1's are equally likely, what is the probability that a 1 was sent given that we received a 1?

4.7. To improve the reliability of the channel described in the last example, we repeat each digit in the message three times. What is the probability that 111 was sent given that (a) we received 101? (b) we received 000?

4.8. Two hunters shoot at a deer, which is hit by exactly one bullet. If the first hunter hits his targets with probability 0.3 and the second with probability 0.6, what is the probability the second hunter killed the deer? The answer is not 2/3. Do you think the answer is larger or smaller?

4.9. A student goes to class on a snowy day with probability 0.4, but on a nonsnowy day attends with probability 0.7. Suppose that 20% of the days in February are snowy. What is the probability it snowed on February 7th given that the student was in class on that day?

4.10. You are about to have an interview for Harvard Law School. 60% of the interviewers are conservative and 40% are liberal. 50% of the conservatives smoke cigars but only 25% of the liberals do. Your interviewer lights up a cigar. What is the probability he is a liberal?

4.11. Five pennies are sitting on a table. One is a trick coin that has Heads on both sides, but the other four are normal. You pick up a penny at random and flip it four times, getting Heads each time. Given this, what is the probability you picked up the two-headed penny?

4.12. One slot machine pays off 1/2 of the time, while another pays off 1/4 of the time. We pick one of the machines and play it six times, winning 3 times. What is the probability we are playing the machine that pays off only 1/4 of the time?

4.13. Returning to Example 2.6, suppose that a person picks 13 cards out of a deck of 52. Let B = "He gets the ace of hearts," A_k = "He gets at least k aces," E_k = "he gets exactly k aces," and $p_k = P(E_k)$. Show that

$$P(E_1|A_1) = \frac{p_1}{p_1 + p_2 + p_3 + p_4} > \frac{p_1}{p_1 + 2p_2 + 3p_3 + 4p_4} = P(E_1|B)$$

so $P(A_2|A_1) < P(A_2|B)$.

4.14. 20% of people are "accident-prone" and have a probability 0.15 of having an accident in a one-year period in contrast to a probability of 0.05 for the other 80% of people. (a) If we pick a person at random, what is the probability he will have an accident this year? (b) What is the probability a person is accident-prone if he had an accident last year? (c) What is the probability he will have an accident this year if he had one last year?

4.15. One die has 4 red and 2 white sides; a second has 2 red and 4 white sides. (a) If we pick a die at random and roll it, what is the probability the result is a red side? (b) If the first result is a red side and we roll the same die again, what is the probability of a second red side?

4.16. Three factories manufacture 20%, 30%, and 50% of the computer chips a company sells. If the fractions of defective chips are 0.4%, 0.3%, and 0.2%, respectively, what fraction of the defective chips come from the third factory?

4.17. In a certain city 30% of the people are Conservatives, 50% are Liberals, and 20% are Independents. In a given election, 2/3 of the Conservatives voted, 80% of the Liberals voted, and 50% of the Independents voted. If we pick a voter at random what is the probability she is Liberal?

4.18. A company gives a test to 100 salesmen, 80 with good sales records and 20 with poor records. 60% of the good salesmen pass the test, but only 30% of the poor salesmen do. A new applicant takes the test and passes. What is the probability he is a good salesman?

4.19. You are a serious student who studies on Friday nights but your roommate goes out and has a good time. 40% of the time he goes out with his girlfriend; 60% of the time he goes to a bar. 30% of the times when he goes out with his girlfriend he spends the night at her apartment. 40% of the times when he goes to a bar he gets in a fight and gets thrown in jail. You wake up on Saturday morning and your roomate is not home. What is the probability he is in jail?

4.20. Two masked robbers try to rob a crowded bank during the lunch hour but the teller presses a button that sets off an alarm and locks the front door. The robbers, realizing they are trapped, throw away their masks and disappear into the chaotic crowd. Confronted with 40 people claiming they are innocent, the police give everyone a lie detector test. Suppose that guilty people are detected with probability 0.95, and innocent people appear to be guilty with probability 0.01. What is the probability Mr. Jones is guilty given that the lie detector says he is?

4.21. Use Bayes formula to show that $P(A|B) > P(A)$ if and only if $P(B|A^c) < P(B|A)$. In words, conditioning on B increases the probability of A if and only if B is more likely when A occurs than when it does not.

*2.5. Recursion

In some cases, it is too complicated to directly compute the probability of some event but it is possible to find the answer we seek by relating it to other probabilities in a suitable family of events. The general description just given will make more sense after the reader sees some concrete examples.

Example 5.1. Gambler's Ruin. Consider a gambling game in which you win \$1 with probability p and lose \$1 with probability $1 - p$, where $0 < p < 1$. Suppose you start with \$50 and decide to quit when you have \$100 or have lost your original \$50. What is the probability you will go home with \$100?

To answer this question we will compute $a_i = $ the probability you reach \$100 before \$0 when you start with \$$i$. Clearly $a_0 = 0$ and $a_{100} = 1$. By considering what happens on the first play, we have

$$(5.1) \qquad a_i = pa_{i+1} + (1 - p)a_{i-1} \qquad \text{for } 1 \le i \le 99$$

The last equation can be written as $pa_{i+1} = a_i - (1 - p)a_{i-1}$. Dividing both sides by p and writing out the first few equations,

$$a_2 = a_1/p - a_0(1 - p)/p$$
$$a_3 = a_2/p - a_1(1 - p)/p$$
$$a_4 = a_3/p - a_2(1 - p)/p$$

it becomes clear that if we knew a_0 and a_1 we could compute $a_2, a_3, a_4, \ldots$

Unfortunately, what we know is a_0 and a_{100}. To get around this difficulty, we will find the general solution of (5.1), or more generally of

$$(5.2) \qquad \alpha a_{i+1} + \beta a_i + \gamma a_{i-1} = 0$$

To find the solutions, we guess $a_i = \lambda^i$. This will be a solution if

$$\alpha \lambda^{i+1} + \beta \lambda^i + \gamma \lambda^{i-1} = 0$$

or, factoring out λ^{i-1}, if

$$(5.3) \qquad \alpha \lambda^2 + \beta \lambda + \gamma = 0$$

Suppose for the moment that (5.3) has two real roots $\lambda_1 \ne \lambda_2$. Equation (5.2) is "linear," that is, if a_i and a_i' are solutions then $a_i + a_i'$ and ca_i are solutions for any real number c. Using the last observation with the two solutions we have found, it follows that

$$(5.4) \qquad c_1 \lambda_1^i + c_2 \lambda_2^i$$

is a solution for any real numbers c_1 and c_2. We claim that by suitable choice of c_1 and c_2, we can achieve any a_0 and a_1. To do this we set

$$a_0 = c_1 + c_2 \qquad a_1 = c_1 \lambda_1 + c_2 \lambda_2$$

and solve. Multiplying the first equation by λ_1 and subtracting it from the second gives $a_1 - \lambda_1 a_0 = (\lambda_2 - \lambda_1)c_2$ so

$$c_2 = \frac{a_1 - \lambda_1 a_0}{\lambda_2 - \lambda_1}$$

Since the first equation implies $c_1 = a_0 - c_2$, we have shown that in the case of two real roots, any choice of a_0 and a_1 can be achieved by suitable choice of c_1 and c_2. Our earlier observation that any solution is determined once we know a_0 and a_1 is valid for (5.2), so (5.4) gives the general solution of (5.2).

The equation we are interested in has $\alpha = p$, $\beta = -1$, and $\gamma = (1-p)$, so we want to find solutions to

$$0 = p\lambda^2 - \lambda + (1-p) = (\lambda - 1)(p\lambda - (1-p))$$

The last computation shows that $\lambda_1 = 1$ and $\lambda_2 = (1-p)/p$ are solutions of (5.3). If $p \neq 1/2$ we have two real roots, so the general solution is

$$c_1 1^i + c_2\{(1-p)/p\}^i$$

To find the particular solution we are interested in we set

$$c_1 + c_2 = a_0 = 0 \qquad c_1 + c_2\{(1-p)/p\}^{100} = a_{100} = 1$$

The first equation implies $c_1 = -c_2$, so substituting this into the second equation we find $c_2 = \left(\{(1-p)/p\}^{100} - 1\right)^{-1}$ and hence

(5.5)
$$a_i = \frac{\left(\frac{1-p}{p}\right)^i - 1}{\left(\frac{1-p}{p}\right)^{100} - 1}$$

To see what this says about gambling, we consider two concrete examples.

Roulette. A roulette wheel has 18 outcomes that are red, 18 that are black, and 2 that are green. If we bet \$1 on black then we win \$1 with probability $p = 18/38 = 0.4736$ and lose \$1 with probability 20/38. In this case

$$a_{50} = \frac{\left(\frac{20}{18}\right)^{50} - 1}{\left(\frac{20}{18}\right)^{100} - 1} = 0.005127$$

or we will succeed in reaching \$100 before \$0 only about 5 times out of 1000.

Craps. As we computed in Example 3.4, the probability of winning at craps is 0.493, so

$$a_{50} = \frac{\left(\frac{0.507}{0.493}\right)^{50} - 1}{\left(\frac{0.507}{0.493}\right)^{100} - 1} = 0.1978$$

EXERCISE 5.1. Show that if $a < x < b$ and we start with x dollars then the probability we reach b before a is

$$\frac{\left(\frac{1-p}{p}\right)^{x} - \left(\frac{1-p}{p}\right)^{a}}{\left(\frac{1-p}{p}\right)^{b} - \left(\frac{1-p}{p}\right)^{a}}$$

EXERCISE 5.2. A roulette player in Las Vegas has $20 but needs $80 to buy a bus ticket back to Minneapolis. What is the probability he will end up with $80 if (a) he bets $20 on red and if he wins bets $40 on red, or (b) he bets $1 every time on red until he reaches $80 or $0?

EXERCISE 5.3. An investor has a stock that each month goes up $1 with probability 0.6 and down $1 with probability 0.4. If she bought the stock when it cost $105 and will sell it when it reaches $200 or falls to $100, what is the probability she will end up with $200?

EXERCISE 5.4. A gambler wants to go home a winner, so he brings $100 to play roulette, betting $1 on red each time. Suppose he quits any time his net winnings are positive. What is the probability he will go home a winner?

EXERCISE 5.5. Generalize the result in the last problem to show that if a gambler with an infinite fortune plays a game with a probability $p < 1/2$ of winning then the probability he will ever be b dollars ahead is $(p/(1-p))^{b}$.

EXERCISE 5.6. A casino owner wins $1 with probability 0.55 and loses $1 with probability 0.45. If he starts with $60, what is the probability he will ever hit $0?

The formulas above are valid only when $p \neq 1/2$. The case of a fair game, $p = 1/2$, is also interesting.

Example 5.2. Two people are flipping a coin. When Heads comes up, George pays Henry $1; when Tails comes up, Henry pays George $1. George has $50 and Henry has $25. What is the probability Henry will go broke first?

We begin by observing that the sum of the two players' fortunes is always \$75, so if George has \$$i$, Henry has \$$(75 - i)$. Let a_i be the probability Henry goes broke first when George starts with \$$i$. Clearly $a_0 = 0$ and $a_{75} = 1$. By considering what happens on the first play, we see that

$$a_i = \frac{1}{2}a_{i+1} + \frac{1}{2}a_{i-1} \qquad \text{for } 1 \le i \le 74$$

Rearranging the last equation gives $a_{i+1} - a_i = a_i - a_{i-1}$, i.e., $a_{i+1} - a_i$ is a constant c that does not depend on i. Since $a_0 = 0$, it follows from the last observation that

$$a_i = a_i - a_0 = (a_i - a_{i-1}) + (a_{i-1} - a_{i-2}) + \cdots + (a_1 - a_0) = ci$$

or, to argue geometrically, since $a_0 = 0$ and a_i has constant slope it must be a straight line through the origin. To have $a_{75} = 1$, we must have $c = 1/75$ so $a_i = i/75$. In words, the probability you will win is equal to the fraction of money in the game that you have.

EXERCISE 5.7. Show that if we are playing a fair game (i.e., $p = 1/2$) and we start with x dollars then the probability we reach $b > x$ before $a < x$ is

(5.6) $$\frac{x - a}{b - a}$$

EXERCISE 5.8. Apply the last result and let $b \to \infty$ to conclude that the probability we reach a starting from x is 1.

For the rest of this section, we will consider recursions that cannot be solved exactly but can be solved numerically with the help of a computer.

Example 5.3. Board games. Suppose you are playing a board game in which you roll a die to determine how many spaces to move your marker forward. Most games have special squares that cause your marker to move forward or backward, but to get a tractable problem we will suppose these features are not present. Suppose that the 34th space along the board carries a reward that you especially like. What is the probability you will hit that square?

Let a_i be the probability that you will hit the square that is i spaces in front of you. By considering what happens on the first roll we have for $i \ge 1$

(5.7) $$a_i = \frac{1}{6}(a_{i-1} + a_{i-2} + a_{i-3} + a_{i-4} + a_{i-5} + a_{i-6})$$

where to make the equation valid for all i we set $a_0 = 1$, and $a_j = 0$ for $j = -1, -2, -3, -4, -5$. In theory one can solve this problem by the methods we used to analyze Gambler's Ruin: $a_i = \lambda^i$ will be a solution of the equation (5.7) if and only if

$$(5.8) \qquad 6\lambda^6 - \lambda^5 - \lambda^4 - \lambda^3 - \lambda^2 - \lambda - 1 = 0$$

and if the equation has six distinct roots λ_i, $i = 1, \ldots, 6$ then the desired solution is found by picking c_i so that $\sum_{i=1}^6 c_i \lambda_i^k$ has the right values at $k = 0, -1, \ldots, -5$. In practice it is much simpler just to use the relationship (5.7) to compute a_1, then a_2, and so on until you have the value you want. The next table gives the values of a_1 to a_6 on the first row, a_7 to a_{12} on the second, and so until we come to a_{42} in the lower right corner.

.166667	.194444	.226852	.264660	.308771	.360232
.253604	.268094	.280369	.289288	.293393	.290830
.279263	.283540	.286114	.287071	.286702	.285587
.284713	.285621	.285968	.285944	.285756	.285598
.285600	.285748	.285769	.285736	.285701	.285692
.285707	.285725	.285722	.285714	.285710	.285712
.285715	.285716	.285715	.285714	.285714	.285714

As the table should indicate, $a_i \to 2/7$ as $i \to \infty$. A proof of this is beyond the scope of this book but it is easy to explain the answer. The "average number of spaces we move on one roll" (this will be explained in Chapter 4) is

$$\frac{1 + 2 + 3 + 4 + 5 + 6}{6} = \frac{21}{6} = \frac{7}{2}$$

So on two rolls we will on the average move 7 spaces, visiting 2 of them and missing the other 5.

EXERCISE 5.9. Suppose we are playing a game in which we flip a coin to determine whether we move forward one space (Heads) or two (Tails). In this case the equation is $a_i = (a_{i-1} + a_{i-2})/2$, with $a_0 = 1$ and $a_{-1} = 0$. Find a formula for a_i.

EXERCISE 5.10. In the game described in the last problem, the only way to miss $i + 1$ is to land on i and then get Tails. Use this to argue that if $\alpha = \lim_{i \to \infty} a_i$ exists then $1 - \alpha = \alpha/2$, so $\alpha = 2/3$.

EXERCISE 5.11. Generalize the last argument to show that if on one turn we move a random number of steps X and if $\alpha = \lim_{i \to \infty} a_i$ exists then $\alpha = 1/\sum_{i=0}^\infty P(X > i)$.

EXERCISE 5.12. Consider a game in which you move forward 1, 2, or 3 spaces with probabilities 1/6, 2/3, and 1/6. Find a formula for a_i in this case.

EXERCISE 5.13. Solve the last problem when the three moves have probability 1/3 each. Note: This requires the use of complex numbers.

Example 5.4. The advantage of going first. Consider a board game of the simple type considered in the last example, which is based on rolling a six-sided die. Suppose that the finish is 50 squares away and suppose that when you are k squares from the finish and you roll a number $\geq k$ you win. (Some games require you to roll exactly k to win in this situation, but again we are trying to keep things simple.) What is the probability that the person who goes first will win?

Let $a(i, j)$ be the probability that the person whose turn it is to roll wins when he has i spaces to go and his opponent has j spaces to go. We want to find $a(50, 50)$. By considering what happens on one roll we have

$$(5.9) \qquad\qquad a(i, j) = 1 - \frac{1}{6} \sum_{k=1}^{6} a(j, i - k)$$

To explain the last equation, suppose that Fred has i steps to go; George has j steps to go; it is Fred's turn and he rolls a 4. After Fred's move, it is George's turn to roll and George has a probability $a(j, i - 4)$ of winning. To complete the description of $a(i, j)$, we note that the game stops when someone reaches the finish, so $a(i, j) = 0$ when $i > 0$ and $j \leq 0$.

There are no analytical results that allow us to determine all the solutions of (5.9), so we have to turn to the computer. We begin by using (5.9) (or common sense) to find that $a(1, 1) = 1$, and then compute the values at

$$(1, 2), (2, 1), \ (1, 3), (2, 2), (3, 1), \ (1, 4), (2, 3), (3, 2), (4, 1), \ldots$$

Note that we have arranged the pairs (i, j) so that $i + j$ is nondecreasing since the sum of the indices on the right-hand side of (5.9) is strictly smaller than $i + j$. A short computer program yields the answer we seek in a few seconds: $a(50, 50) = 0.57596$. Figure 2.1 shows a plot of $a(i, i)$ for $i = 1, \ldots, 100$ and compares the values with

$$0.5 + \frac{7}{4} \sqrt{\frac{3}{10\pi} \frac{1}{\sqrt{i}}}$$

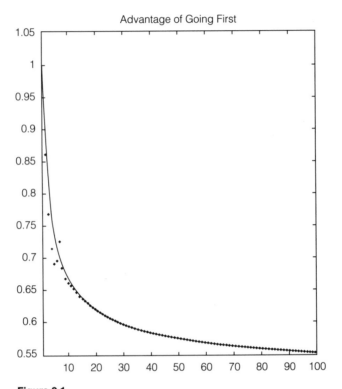

Figure 2.1

We will leave the form of the approximating curve as a little mystery, but observe that it fits the data rather well when i is large. For those who would like to try to figure out the mystery we observe that if T_1 and T_2 are the total numbers of tosses the first and second player need to complete the game, then by symmetry the advantage of going first is $P(T_1 = T_2)/2$. One then gets the answer given above by applying the central limit theorem to $T_1 - T_2$.

EXERCISE 5.14. Solve the last problem when the rules say that if you are k spaces from the finish and roll a number $> k$ you cannot move.

EXERCISE 5.15. Suppose that in a volleyball match, Team A wins a point while serving with probability 0.12 and Team B wins a point while serving with probability 0.15. Compute the probability that A will be the first to score 15 points if they serve first.

EXERCISE 5.16. Real volleyball matches have the feature that you have to win by two points. Suppose the score is tied 14–14 with A serving. What is the probability A will win the match if the two teams have the probabilities given in the last exercise?

Example 5.5. Blackjack. In the game of blackjack a King, Queen, or Jack counts 10, an Ace counts 1 or 11, and the other cards count the numbers that are shown on them (e.g., a 5 counts 5). The object of the game is to get as close to 21 as you can without going over. You start with 2 cards and draw cards out of the deck until either you are happy with your total or you go over 21, in which case you "bust" and you lose. The first step in figuring out how to play blackjack is to consider what is going to happen to the dealer who plays by a fixed rule: He draws a card if his total is ≤ 16, otherwise he stops.

To simplify things, suppose for the moment that an Ace always counts 11. We want to calculate $a(j,k)$ = the probability the dealer's ending total is k when he has a total of j. If $17 \leq j$ then $a(j,j) = 1$ and $a(j,k) = 0$ for $k \neq j$ since in this case the dealer will not take any more cards. We can now compute all the $a(j,k)$ by starting with $j = 16$ and working down to $j = 2$ by using the recursion

$$a(j,k) = \frac{1}{13} \cdot a(j+11,k) + \sum_{m=2}^{9} \frac{1}{13} \cdot a(j+m,k) + \frac{4}{13} \cdot a(j+10,k)$$

Here, the three terms refer to an Ace, 2 through 9, or a card that counts 10 (King, Queen, Jack, or 10) being drawn, and we are simplifying the calculation by assuming that cards are being dealt from an infinite deck in which each value appears on 1/13 of the cards. This assumption is needed to keep the computations from getting extremely complicated, but is not as bad as it might appear. At some casinos, cards are dealt from a "shoe" that contains four or more decks and the decks are shuffled when more than half of the cards have been played.

To deal with the complication that an Ace can count as 1 or 11, we introduce $b(j,k)$ = the probability that the dealer's ending total is k when he has a total of j including one Ace that is being counted as 11. Such hands are called "soft" because even if you draw a 10 you will not bust. With soft hands taken care of in b, we redefine $a(j,k)$ = the probability the dealer's ending total is k when he has a hard total of j, i.e., a hand in which any Ace is counted as 1. Again we start by observing that $a(j,j) = b(j,j) = 1$ when $j \geq 17$ and then start with 16 and work down. However, this time there are several cases to consider. The details of the computation are not very important, so if you find them boring, skip the next paragraph.

To keep the formulas as small as possible we will let $p_i = 1/13$ for $i < 9$

and $p_{10} = 4/13$. If $11 \le j \le 16$ then a new Ace must count as 1 so

$$a(j,k) = p_1 a(j+1,k) + \sum_{m=2}^{10} p_m a(j+m,k)$$

For soft hands, if the card we draw takes us over 21 then we have to change the Ace from counting 11 to counting 1, producing a hard hand, so

$$b(j,k) = p_1 b(j+1,k) + \sum_{m=2}^{21-j} p_m b(j+m,k) + \sum_{m=22-j}^{10} p_m a(j+m-10,k)$$

When $j = 11$ the second sum runs from 11 to 10 and is considered to be 0. Finally, when $2 \le j \le 10$ a new Ace counts as 11 and produces a soft hand:

$$a(j,k) = p_1 b(j+11,k) + \sum_{m=2}^{10} p_m a(j+m,k)$$

Note: By convention (and by casino practice) an Ace counts as 11 if it can, so there are no soft hands with totals of less than 12.

The last three formulas are a little messy but are not hard to manipulate using a computer. The next table gives the probabilities of the various results for the dealer conditional on the value of his first card. We have broken things down this way because when blackjack is played in a casino, we can see one of the dealer's two cards.

	17	18	19	20	21	bust
2	.13981	.13491	.12966	.12403	.11799	**.35361**
3	.13503	.13048	.12558	.12033	.11470	**.37387**
4	.13049	.12594	.12139	.11648	.11123	**.39447**
5	.12225	.12225	.11770	.11315	.10825	**.41640**
6	.16544	.10627	.10627	.10171	.09716	**.42315**
7	**.36857**	.13780	.07863	.07863	.07407	.26231
8	.12857	**.35934**	.12857	.06939	.06939	.24474
9	.12000	.12000	**.35076**	.12000	.06082	.22843
10	.11142	.11142	.11142	**.34219**	.11142	.21211
ace	.13079	.13079	.13079	.13079	**.36156**	.11529

You should note that when the dealer's first card is 2, 3, 4, 5, or 6, she is quite likely to bust, but when her first card is $k = 7, 8, 9, 10$, or Ace $= 11$, her most likely total is $10 + k$. To make this clear we have given the most likely probabilities in boldface.

The analysis of the player's options is even more complicated than that of the dealer's so we will not attempt it here. The first analysis was performed in the mid-'50s (see Baldwin et al. in *J. Amer. Stat. Assoc.* 51, 429–439) and has been redone by a number of other people since that time. To describe the optimal strategy in a few words we use "stand on n" as short for "take a card if your total is $< n$ but not if it is $\geq n$."

Hard Hands

Stand on 17 if the dealer shows 7, 8, 9, 10, or A
Stand on 12 if the dealer shows 2, 3, 4, 5, or 6
Exception: Draw to 12 if the dealer shows 2 or 3

Soft Hands

Stand on 18
Exception: Draw to 18 if the dealer has 9 or 10

To help remember the rules for hard hands, notice that with two exceptions the strategy there is a combination of "mimic the dealer" and "never bust" (that is, "only take a card if you have 11 or less").

Using these rules, the probability you will win is about 0.49, close enough to even if you are only looking for an evening's entertainment. If you have the patience to learn about "splitting pairs" and "doubling down" then you can make blackjack an even game. The necessary details can be found in Chapter 3 of Edward Thorp's book *Beat the Dealer*, which astonished the world in 1962 by demonstrating that by "counting cards" (i.e., by keeping track of the difference between the numbers of cards you have seen that count 10 and those that count 2 through 6) and by adjusting your betting you can make money from blackjack. Before the reader plans a trip to Las Vegas or Atlantic City, we would like to point out that playing this strategy requires hard work, that making money with it requires a lot of capital, and that casinos are allowed to ask you to leave if they think you are playing it.

2.6. Chapter Summary and Review Problems

Section 2.1. Independence. Two events A and B are **independent** intuitively if the occurrence of A has no effect on the probability of occurrence of B, or formally if

$$P(A \cap B) = P(A)P(B)$$

A sequence of events $A_1, \ldots, A_n$ is said to be **pairwise independent** if

$$P(A_i \cap A_j) = P(A_i)P(A_j) \quad \text{whenever} \quad i \neq j$$

and said to be **independent** if for any $i_1 < i_2 < \cdots < i_k$

$$P(A_{i_1} \cap \cdots \cap A_{i_k}) = P(A_{i_1}) \cdots P(A_{i_k})$$

Binomial distribution. The probability of k successes in n independent trials with success probability p is

(1.1)
$$\binom{n}{k} p^k (1-p)^{n-k}$$

Multinomial distribution. If instead of just success and failure there are k possible outcomes with probabilities p_i, $1 \leq i \leq k$, then the probability of getting exactly n_i outcomes of type i in $n = n_1 + \cdots + n_k$ independent trials is

(1.2)
$$\frac{n!}{n_1! \cdots n_k!} p_1^{n_1} \cdots p_k^{n_k}$$

Geometric distribution. If we are waiting for an event of probability p, the number of trials needed, N, has

(1.3)
$$P(N = k) = (1-p)^{k-1} p \qquad \text{for } k = 1, 2, \ldots$$

Section 2.2. Conditional probability. The conditional probability of B given that A has occurred is

(2.1)
$$P(B|A) = P(B \cap A)/P(A)$$

With A fixed this is a probability, i.e., a way of assigning numbers to events B that satisfies the three defining properties, and hence has all the properties that were proved in Section 1.2. Multiplying each side of (2.1) by $P(A)$ gives the **multiplication rule:**

(2.3)
$$P(A \cap B) = P(A)P(B|A)$$

Section 2.3. Two-stage experiments. If $B_1, \ldots, B_n$ is a **partition**, that is, a sequence of disjoint sets with union Ω, then

(3.1)
$$P(A) = \sum_{i=1}^{n} P(A \cap B_i) = \sum_{i=1}^{n} P(B_i)P(A|B_i)$$

Section 2.4. Bayes formula. To compute $P(B_1|A)$ when $P(B_i)$ and $P(A|B_i)$ are given, use the following three-step procedure:

$$P(B_1|A) = \frac{P(B_1 \cap A)}{P(A)} \qquad \text{definition of conditional probability}$$

$$P(B_i \cap A) = P(B_i)P(A|B_i) \qquad \text{multiplication rule}$$

$$P(A) = \sum_{i=1}^{n} P(B_i \cap A)$$

Combining the three steps gives **Bayes formula**:

$$(4.1) \qquad P(B_1|A) = \frac{P(B_1)P(A|B_1)}{\sum_i P(B_i)P(A|B_i)}$$

Section 2.5.   Recursion. In some cases, it is too complicated to directly compute the probability of some event but it is possible to find the answer we seek by the relating it to other probabilities in a suitable family of events. Sometimes the answers can be found explicitly using difference equations, but usually we have to resort to computer programs to perform the computations.

REVIEW PROBLEMS

6.1. Nine children are seated at random in three rows of three desks. Let $A =$ "Al and Bobby sit in the same row," $B =$ "Al and Bobby both sit at one of the four corner desks." Are A and B independent?

6.2. Suppose A and $B \cup C$ are disjoint, B and C are independent, and $P(C) = P(B) = 2P(A)$. Find $P(A)$.

6.3. Construct an example of pairwise independent events with $A \cap B \cap C = \emptyset$.

6.4. How many times should a coin be tossed so that the probability of at least one head is at least 99%?

6.5. Show that the probability of an even number of heads in n tosses is always $1/2$.

6.6. Find $p_k =$ the probability of getting an even number of 6's in k tosses of a fair die.

6.7. Suppose we repeat an experiment with probability $1/4$ of success four times. Compute the probability of 0, 1, 2, 3, and 4 successes and compare with the probability distribution of the number of Aces when we draw 13 cards out of a deck of 52. In which situation do you expect the probability of no Aces to be larger?

6.8. When Al and Bob play tennis, Al wins a set with probability 0.7 while Bob wins with probability 0.3. What is the probability Al will be the first to win (a) two sets, (b) three sets?

6.9. Chevalier de Mere made money betting that he could "roll at least one 6 in four tries." When people got tired of this wager he changed it to "roll at least one double 6 in 24 tries" but then he started losing money. Compute the probabilities of winning these two bets.

6.10. A friend bets you that when rolling two dice you cannot roll a sum of 7 five times before you roll a double 6. What is the probability you will win this bet? (Note that a sum of 7 is six times as likely as a double 6.)

6.11. Three independent events have probabilities 1/4, 1/3, and 1/2. What is the probability exactly one will occur?

6.12. Three missiles are fired at a target. They will hit it with probabilities 0.2, 0.4, and 0.6. Find the probability that the target is hit by (a) three, (b) two, (a) one, (d) no missiles.

6.13. Suppose we roll two dice. What is the probability that the sum is 7 given that neither die showed a 6?

6.14. Three dice are rolled and the sum is 12. What is the probability the first die is 6?

6.15. What is the most likely total for the sum of four dice and what is its probability?

6.16. You and a friend each roll two dice. What is the probability you will both have (a) the same total, (b) the same two numbers?

6.17. Charlie draws five cards out of a deck of 52. If he gets at least three of one suit, he discards the cards not of that suit and then draws until he again has five cards. For example, if he gets three hearts, one club, and one spade, he throws the two nonhearts away and draws two more. What is the probability he will end up with five cards of the same suit?

6.18. Suppose 60% of the people in a town will get exposed to flu in the next month. If you are exposed and not inoculated then the probability of your getting the flu is 80%, but if you are inoculated that probability drops to 15%. Of two executives at Beta Company, one is inoculated and one is not. What is the probability at least one will not get the flu? Assume that the events that determine whether or not they get the flu are independent.

6.19. John takes the bus with probability 0.3 and the subway with probability 0.7. He is late 40% of the time when he takes the bus but only 20% of the time when he takes the subway. What is the probability he is late for work?

6.20. Three bags lie on the table. One has two gold coins, one has two silver coins, and one has one silver and one gold. You pick a bag at random, and pick out one coin. If this coin is gold, what is the probability you picked from the bag with two gold coins?

6.21. 1 out of 1000 births results in fraternal twins; 1 out of 1500 births results in identical twins. Identical twins must be the same sex but the sexes of fraternal twins are independent. If two girls are twins, what is the probability they are fraternal twins?

6.22. Plumber Bob does 40% of the plumbing jobs in a small town. 30% of the people in town are unhappy with their plumbers but 50% of Bob's customers are unhappy with his work. If your neighbor is not happy with his plumber, what is the probability it was Bob?

6.23. Consider the following data on traffic accidents

age group	% of drivers	accident probability
16 to 25	15	.10
26 to 45	35	.04
46 to 65	35	.06
over 65	15	.08

Calculate (a) the probability a randomly chosen driver will have an accident this year, and (b) the probability a driver is between 46 and 65 given that she had an accident.

6.24. A particular football team is known to run 40% of its plays to the left and 60% to the right. When the play goes to the right, the right tackle shifts his stance 80% of the time, but does so only 10% of the time when the play goes to the left. As the team sets up for the play the right tackle shifts his stance. What is the probability that the play will go to the right?

3 Distributions

3.1. Examples, Poisson Approximation to Binomial

In Section 1.1 we defined a random variable to be a real-valued function defined on the sample space. Since then we have seen a number of examples of random variables. Roll two dice and let $X =$ the sum of the two numbers that appear. Flip a coin 10 times and let $X =$ the number of Heads we get. Roll a die until a 4 appears and let $X =$ the number of rolls we need. In these three cases X is a **discrete random variable**. That is, there is a finite or countable sequence of possible values. The distribution of a discrete random variable, called a **discrete distribution**, is described by giving its **probability function**, that is, by giving the value of $P(X = x)$ for all values of x. To illustrate this notion we will give a number of famous examples. In each case, we will only give the values of $P(X = x)$ when $P(X = x) > 0$. The other values we do not mention are 0.

Example 1.1. Hypergeometric distribution. Consider an urn with M red balls and N black balls. If we draw out n balls then the number of red balls we get, R, has

$$P(R = r) = \frac{\binom{M}{r}\binom{N}{n-r}}{\binom{M+N}{n}} \qquad \text{for } r = 0, \dots, n$$

since the denominator gives the number of ways of picking n of the $M + N$ balls and the numerator gives the number of ways of picking r of the M red balls and $n - r$ of the N black balls. (Recall that by convention $C_{n,k} = 0$ if $k > n$ or $k < 0$.)

Example 1.2. Binomial distribution. If we perform an experiment n times and on each trial there is a probability p of success then the number of successes S has

$$P(S = k) = \binom{n}{k} p^k (1 - p)^{n-k} \qquad \text{for } k = 0, \dots, n$$

In words, S has a binomial distribution with parameters n and p, a phrase we will abbreviate as $S = \text{binomial}(n, p)$.

Example 1.3. Geometric distribution. If we repeat an experiment with probability p of success until a success occurs then the number of trials required N has

$$P(N = n) = (1 - p)^{n-1}p \qquad \text{for } n = 1, 2, \ldots$$

In words, N has a geometric distribution with parameter p, a phrase we will abbreviate as $N = \text{geometric}(p)$.

Example 1.4. Poisson distribution. X is said to have a Poisson distribution with parameter λ if

$$P(X = k) = e^{-\lambda}\frac{\lambda^k}{k!} \quad \text{for } k = 0, 1, 2, \ldots$$

Here $\lambda > 0$ is a parameter. To see that this is a probability function we recall

(1.1)
$$e^x = \sum_{k=0}^{\infty} \frac{x^k}{k!}$$

so the proposed probabilities are nonnegative and sum to 1. The Poisson distribution (with $\lambda = 1$) made an unannounced appearance in the analysis of matching in Example 6.2 of Chapter 1. Our next result will explain why the Poisson distribution arises in a number of situations.

(1.2) Poisson approximation to the binomial. Suppose S_n has a binomial distribution with parameters n and p_n. If $p_n \to 0$ and $np_n \to \lambda$ as $n \to \infty$ then

$$P(S_n = k) \to e^{-\lambda}\frac{\lambda^k}{k!}$$

In words, if we have a large number of independent events with small probability then the number that occur has approximately a Poisson distribution. The key to the proof is the following fact, which is of interest in its own right:

(1.3) If $\lambda_n \to \lambda$ then as $n \to \infty$

$$\left(1 - \frac{\lambda_n}{n}\right)^n \to e^{-\lambda}$$

PROOF: L'Hôpital's rule tells us that if f and g are differentiable at 0 and have $f(0) = g(0) = 0$ and $g'(0) \neq 0$ then

$$\lim_{x \to 0} f(x)/g(x) = f'(0)/g'(0)$$

Applying this result to $f(x) = \ln(1 - x)$ and $g(x) = x$, which have $f'(x) = -1/(1 - x)$ and $g'(x) = 1$, we have

$$\lim_{x \to 0} \frac{\ln(1 - x)}{x} = -1$$

To use the last conclusion, we note that

$$\ln\left\{\left(1 - \frac{\lambda_n}{n}\right)^n\right\} = n \ln\left(1 - \frac{\lambda_n}{n}\right)$$
$$= \frac{\ln(1 - \lambda_n/n)}{\lambda_n/n} \cdot \lambda_n \to -1 \cdot \lambda = -\lambda$$

as $n \to \infty$. From this it follows that

$$\left(1 - \frac{\lambda_n}{n}\right)^n = \exp\left(\ln\left\{\left(1 - \frac{\lambda_n}{n}\right)^n\right\}\right) \to e^{-\lambda} \qquad \square$$

PROOF OF (1.2): To prove the result when $k = 0$, we let $\lambda_n = np_n$ and note that

$$P(S_n = 0) = (1 - p_n)^n = \left(1 - \frac{\lambda_n}{n}\right)^n \to e^{-\lambda}$$

by (1.3). To prove the result for $k > 0$, we observe that

$$P(S_n = k) = \binom{n}{k}\left(\frac{\lambda_n}{n}\right)^k\left(1 - \frac{\lambda_n}{n}\right)^{n-k}$$
$$= \frac{n(n-1)\cdots(n-k+1)}{n^k} \frac{\lambda_n^k}{k!}\left(1 - \frac{\lambda_n}{n}\right)^n\left(1 - \frac{\lambda_n}{n}\right)^{-k}$$
$$\to 1 \cdot \frac{\lambda^k}{k!} \cdot e^{-\lambda} \cdot 1$$

Here $n(n-1)\cdots(n-k+1)/n^k \to 1$ since there are k factors in the numerator and for each fixed j, $(n-j)/n = 1 - (j/n) \to 1$. The last term $(1 - \{\lambda_n/n\})^{-k} \to 1$ since k is fixed and $1 - \{\lambda_n/n\} \to 1$. $\qquad \square$

When we apply (1.2) we think, "If $S_n = \text{binomial}(n, p)$ and p is small then S_n is approximately Poisson(np)." The next example illustrates the use of this approximation and shows that the number of trials does not have to be very large for us to get accurate answers.

Example 1.5. Suppose we roll two dice 12 times and we let D be the number of times a double 6 appears. Here $n = 12$ and $p = 1/36$, so $np = 1/3$. We will now compare $P(D = k)$ with the Poisson approximation for $k = 0, 1, 2$.

$k = 0$ exact answer:

$$P(D = 0) = \left(1 - \frac{1}{36}\right)^{12} = 0.7132$$

Poisson approximation: $\quad P(D = 0) = e^{-1/3} = 0.7165$

$k = 1$ exact answer:

$$P(D = 1) = \binom{12}{1} \frac{1}{36} \left(1 - \frac{1}{36}\right)^{11}$$

$$= \left(1 - \frac{1}{36}\right)^{11} \cdot \frac{1}{3} = 0.2445$$

Poisson approximation: $\quad P(D = 1) = e^{-1/3} \frac{1}{3} = 0.2388$

$k = 2$ exact answer:

$$P(D = 2) = \binom{12}{2} \left(\frac{1}{36}\right)^2 \left(1 - \frac{1}{36}\right)^{10}$$

$$= \left(1 - \frac{1}{36}\right)^{10} \cdot \frac{12 \cdot 11}{36^2} \cdot \frac{1}{2!} = 0.0384$$

Poisson approximation: $\quad P(D = 2) = e^{-1/3} \left(\frac{1}{3}\right)^2 \frac{1}{2!} = 0.0398$

The Poisson distribution is often used as a model for the number of typos per page in a book, the number of defects in a large roll of magnetic tape, or the number of traffic accidents in a day. To explain the reasoning in the last case we note that any one person has a small probability of having an accident on a given day, and it is reasonable to assume that the events $A_i =$ "The ith person has an accident" are independent. Now it is not reasonable to assume that the probabilities of having an accident $p_i = P(A_i)$ are all the same, but fortunately the Poisson approximation does not require this.

(1.4) General Poisson approximation result. Consider independent events A_i, $i = 1, 2, \ldots, n$ with probabilities $p_i = P(A_i)$. Let N be the number of events that occur, let $\lambda = p_1 + \cdots + p_n$, and let Z have a Poisson distribution with parameter λ. For any set of integers B,

$$|P(N \in B) - P(Z \in B)| \le \sum_{i=1}^{n} p_i^2 \le \lambda \max_i p_i$$

This result says that if all the p_i are small then the distribution of N is close to a Poisson with parameter λ. Taking $B = \{k\}$ we see that the individual probabilities $P(N = k)$ are close to $P(Z = k)$, but this result says more. The probabilities of events such as $P(N \ge 3)$ are close to $P(Z \ge 3)$ and we have an explicit bound on the error.

EXERCISES

1.1. Suppose we roll two dice and let X and Y be the two numbers that appear. Find the probability function for $|X - Y|$.

1.2. Suppose we roll three tetrahedral dice that have 1, 2, 3, and 4 on their four sides. Find the probability function for the sum of the three numbers.

1.3. Suppose we draw three balls out of an urn that contains fifteen balls numbered from 1 to 15. Let X be the largest number drawn. (a) Find the probability function for X. (b) What value has the largest probability?

1.4. Suppose X has an exponential distribution, and let $Y = [X]$ be the largest integer $\le X$. Find the probability function for Y.

1.5. What value of p maximizes $P(X = k)$ when X is binomial with parameters n and p? The reason for interest in this question is that it provides the **maximum likelihood estimate** for p when we observe k successes in n trials.

1.6. Suppose X has a Poisson distribution with parameter λ. Compute $P(X = i)/P(X = (i - 1))$ for $i \ge 1$. This recursion is useful for computing Poisson probabilities on a computer and it also shows us that $P(X = i) > P(X = i - 1)$ if and only if $\lambda > i$.

1.7. Compare the Poisson approximation with the exact binomial probabilities when (a) $n = 10$, $p = 0.1$, (b) $n = 20$, $p = 0.05$, (c) $n = 40$, $p = 0.025$.

1.8. The probability of a three of a kind in poker is approximately 1/50. Use the Poisson approximation to compute the probability you will get at least one three of a kind if you play 20 hands of poker.

1.9. In one of the New York state lottery games, a number is chosen at random between 0 and 999. Suppose you play this game 250 times. Use the Poisson approximation to estimate the probability you will never win and compare this with the exact answer.

1.10. Suppose that the probability of a defect in a foot of magnetic tape is 0.002. Use the Poisson approximation to compute the probability that a 1500 foot roll will have no defects.

1.11. Suppose 1% of a certain brand of Christmas lights is defective. Use the Poisson approximation to compute the probability that in a box of 25 there will be at most one defective bulb.

1.12. An airline company sells 200 tickets for a plane with 198 seats, knowing that the probability a passenger will not show up for the flight is 0.01. Use the Poisson approximation to compute the probability they will have enough seats for all the passengers who show up.

1.13. If you bet $1 on number 13 at roulette (or on any other number) then you win $35 if that number comes up, an event of probability 1/38, and you lose your dollar otherwise. (a) Use the Poisson approximation to show that if you play 70 times then the probability you will have won more money than you have lost is larger than 1/2. (b) What is the probability you have lost $70? (c) Won $2?

1.14. An insurance company insures 3000 people, each of whom has a 1/1000 chance of an accident in one year. Use the Poisson approximation to compute the probability there will be at most 2 accidents.

1.15. **Poisson approximation for dependent events.** Suppose for each n we have events $E_1^n, \ldots, E_n^n$ so that for any fixed k,

$$(\star) \qquad \sum_{i_1 < \cdots < i_k} P(E_{i_1}^n \cap \cdots \cap E_{i_k}^n) \to \lambda^k / k!$$

as $n \to \infty$. Let N_n be the number of events that occur. Use the Bonferroni inequalities to conclude that

$$\limsup_{n \to \infty} P(N_n \geq 1) \leq \sum_{j=1}^k (-1)^{j+1} \frac{\lambda^j}{j!} \quad \text{for odd } k$$

$$\liminf_{n \to \infty} P(N_n \geq 1) \geq \sum_{j=1}^k (-1)^{j+1} \frac{\lambda^j}{j!} \quad \text{for even } k$$

so $P(N_n = 0) \to e^{-\lambda}$. By working harder one can show that under this assumption, $P(N_n = k) \to e^{-\lambda} \lambda^k / k!$.

1.16. Show that $(\star)$ is satisfied in the Matching Problem, Example 6.2 in Chapter 1.

1.17. Show that if we consider the Birthday Problem, Example 5.1 in Chapter 1, for a calendar with m days and we have $n = \sqrt{\lambda m}$ people then $(\star)$ holds.

3.2. Density and Distribution Functions

In many situations random variables can take any value on the real line or in a certain subset of the real line. For concrete examples, consider the height or weight of a person chosen at random or the time it takes a person to drive from Los Angeles to San Francisco. A random variable X is said to have a **continuous distribution** with **density function** f if for all $a \leq b$ we have

$$(2.1) \qquad P(a \leq X \leq b) = \int_a^b f(x)\,dx$$

Geometrically, $P(a \leq X \leq b)$ is the area under the curve f between a and b.

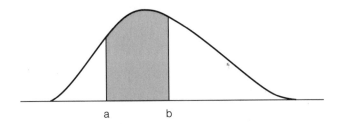

$$a \qquad b$$

    in For the purposes of understanding and remembering formulas, it is useful to think of $f(x)$ as $P(X = x)$ even though the last event has probability zero. To explain the last remark and to prove $P(X = x) = 0$, note that taking $a = x$ and $b = x + \Delta x$ in (2.1) we have

$$P(x \leq X \leq x + \Delta x) = \int_x^{x+\Delta x} f(y)\,dy \approx f(x)\Delta x$$

when Δx is small. Letting $\Delta x \to 0$, we see that $P(X = x) = 0$, but $f(x)$ tells us how likely it is for X to be near x. That is,

$$\frac{P(x \leq X \leq x + \Delta x)}{\Delta x} \approx f(x)$$

    In order for $P(a \leq X \leq b)$ to be nonnegative for all a and b and for $P(-\infty < X < \infty) = 1$ we must have

$$(2.2) \qquad f(x) \geq 0 \quad \text{and} \quad \int_{-\infty}^{\infty} f(x)\,dx = 1$$

Any function f that satisfies (2.2) is said to be a **density function.** Some important density functions are:

Example 2.1. The uniform distribution on (a,b).

$$f(x) = \begin{cases} \frac{1}{b-a} & a < x < b \\ 0 & \text{otherwise} \end{cases}$$

The idea here is that we are picking a value "at random" from (a,b). That is, values outside the interval are impossible, and all those inside have the same probability (density). If we set $f(x) = c$ when $a < x < b$ and 0 otherwise then

$$\int f(x)\,dx = \int_a^b c\,dx = c(b-a)$$

So we have to pick $c = 1/(b-a)$ to make the integral 1. The most important special case occurs when $a = 0$ and $b = 1$. Random numbers generated by a computer are typically uniformly distributed on (0,1). Another case that comes up in applications is $a = -1/2$ and $b = 1/2$. If we take a measurement and round it off to the nearest integer then it is reasonable to assume that the "round-off error" is uniformly distributed on $(-1/2, 1/2)$.

Example 2.2. Power laws.

$$f(x) = \begin{cases} (\rho-1)x^{-\rho} & x \geq 1 \\ 0 & \text{otherwise} \end{cases}$$

Here $\rho > 1$ is a parameter that governs how fast the probabilities go to 0 at ∞. To check that this is a density function, we note that

$$\int_1^\infty (\rho-1)x^{-\rho}\,dx = -x^{-(\rho-1)}\Big|_1^\infty = 0 - (-1) = 1$$

These distributions are often used in situations where $P(X > x)$ does not go to 0 very fast as $x \to \infty$. For example, the Italian economist Pareto used them to describe the distribution of family incomes.

Example 2.3. The exponential distribution.

$$f(x) = \begin{cases} \lambda e^{-\lambda x} & x \geq 0 \\ 0 & \text{otherwise} \end{cases}$$

Here $\lambda > 0$ is a parameter. To check that this is a density function, we note that

$$\int_0^\infty \lambda e^{-\lambda x}\,dx = -e^{-\lambda x}\Big|_0^\infty = 0 - (-1) = 1$$

Exponentially distributed random variables often come up as waiting times between events; for example, the arrival times of customers at a bank or ice cream shop. Sometimes we will indicate that X has an exponential distribution with parameter λ by writing $X = \text{exponential}(\lambda)$.

Last and perhaps most important is

Example 2.4. The standard normal distribution.

$$f(x) = (2\pi)^{-1/2} e^{-x^2/2}$$

Since there is no closed form expression for the antiderivative of f, it takes some ingenuity to check that this is a probability density. Let $I = \int e^{-x^2/2}\, dx$. To show that $\int f(x)\, dx = 1$, we want to show that $I = \sqrt{2\pi}$.

$$I^2 = \int e^{-x^2/2}\, dx \int e^{-y^2/2}\, dy = \iint e^{-(x^2+y^2)/2}\, dx\, dy$$

Changing to polar coordinates, the last integral becomes

$$\int_0^\infty \int_0^{2\pi} e^{-r^2/2}\, r\, d\theta\, dr = 2\pi \int_0^\infty e^{-r^2/2}\, r\, dr = 2\pi \left(-e^{-r^2/2}\right)\Big|_0^\infty = 2\pi$$

So $I^2 = 2\pi$ or $I = \sqrt{2\pi}$.

Any random variable (discrete, continuous, or in between) has a **distribution function** defined by $F(x) = P(X \le x)$. If X has a density function $f(x)$ then

$$F(x) = P(-\infty < X \le x) = \int_{-\infty}^x f(y)\, dy$$

That is, F is an antiderivative of f.

One of the reasons for computing the distribution function is explained by the next formula. If $a < b$ then $\{X \le b\} = \{X \le a\} \cup \{a < X \le b\}$ with the two sets on the right-hand side disjoint so

$$P(X \le b) = P(X \le a) + P(a < X \le b)$$

or, rearranging,

(2.3) $P(a < X \le b) = P(X \le b) - P(X \le a) = F(b) - F(a)$

The last formula is valid for any random variable. When X has density function f, it says that

$$\int_a^b f(x)\, dx = F(b) - F(a)$$

i.e., the integral can be evaluated by taking the difference of the antiderivative at the two endpoints.

To see what distribution functions look like, and to explain the use of (2.3), we return to our examples.

Example 2.5. The uniform distribution. $f(x) = 1/(b-a)$ for $a < x < b$.

$$F(x) = \begin{cases} 0 & x \leq a \\ (x-a)/(b-a) & a \leq x \leq b \\ 1 & x \geq b \end{cases}$$

To check this, note that $P(a < X < b) = 1$ so $P(X \leq x) = 1$ when $x \geq b$ and $P(X \leq x) = 0$ when $x \leq a$. For $a \leq x \leq b$ we compute

$$P(X \leq x) = \int_{-\infty}^{x} f(y)\,dy = \int_{a}^{x} \frac{1}{b-a}\,dy = \frac{x-a}{b-a}$$

In the most important special case $a = 0$, $b = 1$ we have $F(x) = x$ for $0 \leq x \leq 1$.

Example 2.6. The power laws. $f(x) = (\rho - 1)x^{-\rho}$ for $x \geq 1$. Here $\rho > 1$.

$$F(x) = \begin{cases} 0 & x \leq 1 \\ 1 - x^{-(\rho-1)} & x \geq 1 \end{cases}$$

The first line of the answer is easy to see. Since $P(X > 1) = 1$, we have $P(X \leq x) = 0$ for $x \leq 1$. For $x \geq 1$ we compute

$$P(X \leq x) = \int_{-\infty}^{x} f(y)\,dy = \int_{1}^{x} (\rho - 1)y^{-\rho}\,dy$$

$$= -y^{-(\rho-1)}\Big|_{1}^{x} = 1 - x^{-(\rho-1)}$$

To illustrate the use of (2.3) we note that if $\rho = 3$ then

$$P(2 < X \leq 4) = (1 - 4^{-2}) - (1 - 2^{-2}) = \frac{1}{4} - \frac{1}{16} = \frac{3}{16}$$

Example 2.7. The exponential distribution. $f(x) = \lambda e^{-\lambda x}$ for $x \geq 0$.

$$F(x) = \begin{cases} 0 & x \leq 0 \\ 1 - e^{-\lambda x} & x \geq 0 \end{cases}$$

The first line of the answer is easy to see. Since $P(X > 0) = 1$ we have $P(X \leq x) = 0$ for $x \leq 0$. For $x \geq 0$ we compute

$$P(X \leq x) = \int_{-\infty}^{x} f(y)\,dy = \int_{0}^{x} \lambda e^{-\lambda y}\,dy$$

$$= -e^{-\lambda y}\Big|_{0}^{x} = -e^{-\lambda x} - (-1)$$

Suppose X has an exponential distribution with parameter λ. If $t \geq 0$ then $P(X > t) = 1 - P(X \leq t) = 1 - F(t) = e^{-\lambda t}$, so if $s \geq 0$ then

$$P(T > t + s | T > t) = \frac{P(T > t + s)}{P(T > t)} = \frac{e^{-\lambda(t+s)}}{e^{-\lambda t}} = e^{-\lambda s} = P(T > s)$$

This is the **lack of memory property** of the exponential distribution. Given that you have been waiting t units of time, the probability you must wait an additional s units of time is the same as if you had not been waiting at all.

Example 2.8. The standard normal distribution. In this case, there is, unfortunately, no closed form expression for the distribution function, so we have to use a table like the one given in the Appendix. Letting Φ denote the normal distribution function, we have

$$P(1 < X \leq 2) = \Phi(2) - \Phi(1) = 0.9772 - 0.8413 = 0.1359$$

The table only gives the values of $\Phi(x)$ for $x \geq 0$. Values for $x < 0$ are computed by noting that the normal density function is symmetric about 0 $(f(x) = f(-x))$ so

$$P(X \leq -x) = P(X \geq x) = 1 - P(X \leq x)$$

since $P(X = x) = 0$. For an example of the use of symmetry, we note that

$$P(X \leq -1) = 1 - P(X \leq 1) = 1 - 0.8413 = 0.1587$$

so

$$P(-1 \leq X \leq 1) = \Phi(1) - \Phi(-1) = 0.8413 - 0.1587 = 0.6826$$

Note that in the general formula in (2.3) it was important to keep track of the difference between $<$ and $\leq$. Here it is not, since for the normal distribution $P(X = x) = 0$ for all x.

EXERCISE 2.1. Suppose X has a normal distribution. Use the table to compute (a) $P(-1 < X < 3)$, (b) $P(X > -2)$, (c) $P(-1.5 < X < -0.5)$.

Distribution functions are somewhat messier in the discrete case.

Example 2.9. Flip three coins and let X be the number of heads that we see. The probability function is given by

x	0	1	2	3
$P(X = x)$	1/8	3/8	3/8	1/8

and the distribution function by (see Figure 3.2)

$$F(x) = \begin{cases} 0 & x < 0 \\ 1/8 & 0 \le x < 1 \\ 4/8 & 1 \le x < 2 \\ 7/8 & 2 \le x < 3 \\ 1 & 3 \le x \end{cases}$$

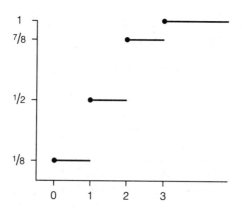

Figure 3.2

To check this, note for example that for $1 \le x < 2$, $P(X \le x) = P(X \in \{0,1\}) = 1/8 + 3/8$. The reader should note that F is discontinuous at each possible value of X and the height of the jump there is $P(X = x)$.

All distribution functions have the following properties

(2.4) If $x_1 < x_2$ then $F(x_1) \le F(x_2)$ i.e., F is nondecreasing.

(2.5) $\lim_{x \to -\infty} F(x) = 0$

(2.6) $\lim_{x \to \infty} F(x) = 1$

(2.7) $\lim_{y \downarrow x} F(y) = F(x)$, i.e., F is continuous from the right.

(2.8) $\lim_{y \uparrow x} F(y) = P(X < x)$

(2.9) $\lim_{y \downarrow x} F(y) - \lim_{y \uparrow x} F(y) = P(X = x)$,
i.e., the jump in F at x is equal to $P(X = x)$.

PROOF: To prove (2.4) we note that $\{X \leq x_1\} \subset \{X \leq x_2\}$, so (2.3) in Chapter 1 implies $F(x_1) = P(X \leq x_1) \leq P(X \leq x_2) = F(x_2)$.

For (2.5), we note that $\{X \leq x\} \downarrow \emptyset$ as $x \downarrow -\infty$ (here $\downarrow$ is short for "decreases and converges to"), so (2.7) from Chapter 1 implies that $P(X \leq x) \downarrow P(\emptyset) = 0$.

The argument for (2.6) is similar $\{X \leq x\} \uparrow \Omega$ as $x \uparrow \infty$ (here $\uparrow$ is short for "increases and converges to"), so (2.6) from Chapter 1 implies that $P(X \leq x) \uparrow P(\Omega) = 1$.

For (2.7), we note that if $y \downarrow x$ then $\{X \leq y\} \downarrow \{X \leq x\}$, so (2.7) from Chapter 1 implies that $P(X \leq y) \downarrow P(X \leq x)$.

The argument for (2.8) is similar. If $y \uparrow x$ then $\{X \leq y\} \uparrow \{X < x\}$ since $\{X = x\} \not\subset \{X \leq y\}$ when $y < x$. Using (2.6) from Chapter 1 now, (2.8) follows.

Subtracting (2.8) from (2.7) gives (2.9).   □

Medians. Intuitively, the median is the place where $F(x)$ crosses $1/2$. The precise definition we are about to give is complicated by the fact that $\{x : F(x) = 1/2\}$ may be empty or contain more than one point.

> m is a **median** for F if $P(X \leq m) \geq 1/2$ and $P(X \geq m) \geq 1/2$.

We begin with a simple example and then consider two that illustrate the problems that can arise.

Example 2.10. Suppose X has an exponential(λ) density. As we computed in Example 2.7, the distribution function is $F(x) = 1 - e^{-\lambda x}$. To find the median we set $P(X \leq m) = 1/2$, i.e., $1 - e^{-\lambda m} = 1/2$, and solve to find $m = (\ln 2)/\lambda$. To see that this is a median, we note that $P(X = m) = 0$ so $P(X \geq m) = P(X > m) = 1 - P(X \leq m) = 1/2$. To see that this is the only median, we observe that if $m < (\ln 2)/\lambda$ then $P(X \leq m) < 1/2$ while if $m > (\ln 2)/\lambda$ then $P(X \geq m) < 1/2$.

In the context of radioactive decay, which is commonly modeled with an exponential distribution, the median is sometimes called the **half-life**, since half of the particles will have broken down by that time. One reason for interest in the half-life is that

$$P(X > k \ln 2/\lambda) = e^{-k \ln 2} = 2^{-k}$$

or in words, after k half-lives only $1/2^k$ particles remain radioactive.

Example 2.11. Suppose X takes values 1, 2, 3 with probability 1/3 each. The distribution function is

$$F(x) = \begin{cases} 0 & x < 1 \\ 1/3 & 1 \le x < 2 \\ 2/3 & 2 \le x < 3 \\ 1 & 3 \le x \end{cases}$$

To check that 2 is a median, we note that

$$P(X \le 2) = P(X \in \{1,2\}) = 2/3$$
$$P(X \ge 2) = P(X \in \{2,3\}) = 2/3$$

This is the only median, since if $x < 2$ then $P(X \le x) \le P(X < 2) \le 1/3$ and if $x > 2$ then $P(X \ge x) \le P(X > 2) = 1/3$.

Example 2.12. Suppose X takes values 1, 2, 3, 4 with probability 1/4 each. The distribution function is

$$F(x) = \begin{cases} 0 & x < 1 \\ 1/4 & 1 \le x < 2 \\ 2/4 & 2 \le x < 3 \\ 3/4 & 3 \le x < 4 \\ 1 & 4 \le x \end{cases}$$

To check that any number m with $2 \le m \le 3$ is a median, we note that for any of these values
$$P(X \le m) \ge P(X \in \{1,2\}) = 2/4$$
$$P(X \ge m) \ge P(X \in \{3,4\}) = 2/4$$

These are the only medians, since if $x < 2$ then $P(X \le x) \le P(X < 2) \le 1/4$ and if $x > 3$ then $P(X \ge x) \le P(X > 3) = 1/4$.

The amount of work involved in computing the median can be reduced by proving

EXERCISE 2.2. Let $m_0 = \min\{x : P(X \le x) \ge 1/2\}$ and $m_1 = \max\{x : P(X \ge x) \ge 1/2\}$. Then $m_0 \le m_1$ and m is a median if and only if $m_0 \le m \le m_1$.

EXERCISES

2.3. Suppose X has a continuous distribution and has $P(1 < X < 2) = 0$. What can we conclude about the density function $f(x)$ and the distribution function $F(x)$?

2.4. $F(x) = 3x^2 - 2x^3$ for $0 < x < 1$ (with $F(x) = 0$ if $x \leq 0$ and $F(x) = 1$ if $x \geq 1$) defines a distribution function. Find the corresponding density function.

2.5. Let $F(x) = e^{-1/x}$ for $x \geq 0$, $F(x) = 0$ for $x \leq 0$. Is F a distribution function? If so, find its density function.

2.6. Let $F(x) = 3x - 2x^2$ for $0 \leq x \leq 1$, $F(x) = 0$ for $x \leq 0$, and $F(x) = 1$ for $x \geq 1$. Is F a distribution function? If so, find its density function.

2.7. Suppose X has density function $f(x) = c(3 - |x|)$ when $-3 < x < 3$. What value of c makes this a density function?

2.8. Consider $f(x) = c(1 - x^2)$ for $-1 < x < 1$, 0 otherwise. What value of c should we take to make f a density function?

2.9. Consider $f(x) = cx^{-1/2}$ for $x \geq 1$, 0 otherwise. Show that there is no value of c that makes this a density function.

2.10. Suppose X has density function $f(x) = x/2$ for $0 < x < 2$, 0 otherwise. Find (a) the distribution function, (b) $P(X < 1)$, (c) $P(X > 3/2)$.

2.11. Suppose X has density function $f(x) = 4x^3$ for $0 < x < 1$, 0 otherwise. Find (a) the distribution function, (b) $P(X < 1/2)$, (c) $P(1/3 < X < 2/3)$.

2.12. Suppose X has density function $x^{-1/2}/2$ for $0 < x < 1$, 0 otherwise. Find (a) the distribution function, (b) the median.

2.13. Suppose X has an exponential distribution with parameter λ. Find $P(X > 2/\lambda)$.

2.14. Suppose $P(X = x) = x/21$ for $x = 1, 2, 3, 4, 5, 6$. Find all the medians of this distribution.

2.15. Suppose X has a Poisson distribution with $\lambda = \ln 2$. Find all the medians of X.

2.16. When $y \geq x$ we have $(1 - 3y^{-4})e^{-y^2/2} \leq e^{-y^2/2} \leq e^{-x^2/2}e^{-x(y-x)}$. Integrate these inequalities from x to ∞ to conclude that

$$(x^{-1} - x^{-3})e^{-x^2/2} \leq 1 - \Phi(x) \leq x^{-1}e^{-x^2/2}$$

This gives accurate bounds on the normal distribution function for large values of x. Compute the upper and lower bounds when $x = 6$.

2.17. Suppose we repeat an experiment with probability p of success and let M be the number of failures we have before the first success. Show that $P(M > m + n | M > m) = P(M > n)$. The geometric distribution we have defined does not have this lack of memory property.

2.18. Let X_n have a geometric distribution with $p = \lambda/n$. Compute $P(X_n/n > x)$ and show that as $n \to \infty$ it converges to $P(Y > x)$ where Y has an exponential distribution with parameter λ.

2.19. Let $T \geq 0$ have density function $f(x)$ and distribution function $F(x)$. The **hazard rate** at time t is the failure rate conditioned on survival up to time t:

$$\lambda(t) = \lim_{h \to 0} \frac{1}{h} P(t < T < t + h | T > t)$$

(a) Express $\lambda(t)$ in terms of f and F. Compute $\lambda(t)$ when (b) $F(x) = 1 - e^{-\lambda x}$, (c) $F(x) = 1 - e^{-x^2}$.

3.3. Functions of Random Variables

In this section we will answer the question: If X has density function f and $Y = r(X)$, then what is the density function for Y? Before proving a general result, we will consider an example:

Example 3.1. Suppose X has an exponential distribution with parameter λ. What is the distribution of $Y = X^2$? To solve this problem we will use the distribution function. First we recall from Example 2.7 that $P(X \leq x) = 1 - e^{-\lambda x}$ so if $y \geq 0$ then

$$P(Y \leq y) = P(X^2 \leq y) = P(X \leq \sqrt{y}) = 1 - e^{-\lambda y^{1/2}}$$

Differentiating, we see that the density function of Y

$$f_Y(y) = \frac{d}{dy} P(Y \leq y) = \frac{\lambda y^{-1/2}}{2} e^{-\lambda y^{1/2}} \qquad \text{for } y \geq 0$$

and 0 otherwise.

Generalizing from the last example, we get

(3.1) Suppose X has density f and $P(a < X < b) = 1$. Let $Y = r(X)$. Suppose $r : (a, b) \to (\alpha, \beta)$ is continuous and strictly increasing, and let $s : (\alpha, \beta) \to (a, b)$ be the inverse of r. Then Y has density

$$g(y) = f(s(y))s'(y) \qquad \text{for } y \in (\alpha, \beta)$$

Before proving this, let's see how it applies to the last example. There X has density $f(x) = \lambda e^{-\lambda x}$ for $x \geq 0$ so we can take $a = 0$ and $b = \infty$. The function $r(x) = x^2$ is indeed continuous and strictly increasing on $(0, \infty)$. (Notice, however, that x^2 is decreasing on $(-\infty, 0)$.) $\alpha = r(a) = 0$ and $\beta = r(b) = \infty$. To find the inverse function we set $y = x^2$ and solve to get $x = y^{1/2}$ so $s(y) = y^{1/2}$. Differentiating, we have $s'(y) = y^{-1/2}/2$ and plugging into the formula, we have

$$g(y) = \lambda e^{-\lambda y^{1/2}} \cdot y^{-1/2}/2 \qquad \text{for } y > 0$$

PROOF: If $y \in (\alpha, \beta)$ then

$$P(Y \le y) = P(r(X) \le y) = P(X \le s(y))$$

since r is increasing and s is its inverse.

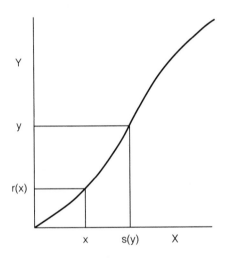

Writing $F(x)$ for $P(X \le x)$ and differentiating with respect to y now gives

$$g(y) = \frac{d}{dy}P(Y \le y) = \frac{d}{dy}F(s(y)) = F'(s(y))s'(y) = f(s(y))s'(y)$$

by the chain rule. $\square$

The next three examples illustrate the use of (3.1). The first is primarily an excuse to make a definition.

Example 3.2. Suppose X has density function $f(x) = (2\pi)^{-1/2}e^{-x^2/2}$, which we called the **standard normal distribution** to distinguish it from the ones we are about to create. Let $Y = \sigma X + \mu$ where $\sigma > 0$. The inverse of $r(x) = \sigma x + \mu$ is $s(y) = (y - \mu)/\sigma$, so (3.1) implies that Y has density function

$$f(s(y))s'(y) = (2\pi)^{-1/2}e^{-\{(y-\mu)/\sigma\}^2/2}\frac{1}{\sigma}$$

$$= (2\pi\sigma^2)^{-1/2}e^{-(y-\mu)^2/2\sigma^2}$$

Y is said to have a **normal distribution with parameters** μ **and** σ^2. We will see what these parameters mean in Chapter 4. For the moment, it is enough to observe that if $Y = \text{normal}(\mu, \sigma^2)$ then reversing the formula we used to define Y, $X = (Y - \mu)/\sigma$ has the standard normal distribution.

For a concrete example of reducing a general normal to a standard one, suppose we are told that a man's height has a normal distribution with $\mu = 69$ inches and $\sigma^2 = 9$ inches. What is the probability a man is more than 6 feet tall (72 inches)? The first step in the solution is to rephrase the question in terms of X

$$P(Y \geq 72) = P(Y - 69 \geq 3) = P\left(\frac{Y - 69}{\sqrt{9}} \geq 1\right)$$

$$= P(X \geq 1) = 1 - 0.8413 = 0.1587$$

from the table in the Appendix.

Example 3.3. How not to water your lawn. The head of a lawn sprinkler, which is a metal rod with a line of small holes in it, revolves back and forth so that drops of water shoot out at angles between 0 and $\pi/2$ radians (i.e., between 0 and 90 degrees). If we use x to denote the distance from the sprinkler and y the height off the ground, then a drop of water released at angle θ with velocity v_0 will follow a trajectory

$$x(t) = (v_0 \cos \theta)t \qquad y(t) = (v_0 \sin \theta)t - gt^2/2$$

where g is the gravitational constant, 32 ft/sec^2. The drop lands when $y(t_0) = 0$ that is, at time $t_0 = (2v_0 \sin \theta)/g$. At this time

$$x(t_0) = \frac{2v_0^2}{g} \sin \theta \cos \theta = \frac{v_0^2}{g} \sin(2\theta)$$

If we assume that the sprinkler moves evenly back and forth between 0 and $\pi/2$, it will spend an equal amount of time at each angle. Letting $K = v_0^2/g$, this leads us to the following question:

If Θ is uniform on $[0, \pi/2]$ then what is the distribution of $Z = K \sin(2\Theta)$?

The first difficulty we must confront when solving this problem is that $\sin(2x)$ is increasing on $[0, \pi/4]$ and decreasing on $[\pi/4, \pi/2]$. The solution to this problem is simple, however. The function $\sin(2x)$ is symmetric about $\pi/4$, so if we let X be uniform on $[0, \pi/4]$ then $Z = K\sin(2\Theta)$ and $Y = K\sin(2X)$ have the same distribution. To apply (3.1) we let $r(x) = K\sin(2x)$ and solve $y = K\sin(2x)$ to get $s(y) = (1/2)\sin^{-1}(y/K)$. Plugging into (3.1) and recalling

$$\frac{d}{dx}\sin^{-1}(x) = \frac{1}{\sqrt{1 - x^2}}$$

we see that Y has density function

$$f(s(y))s'(y) = \frac{4}{\pi} \cdot \frac{1}{2\sqrt{1 - y^2/K^2}} \cdot \frac{1}{K} = \frac{2}{\pi\sqrt{K^2 - y^2}}$$

when $0 < y < K$ and 0 otherwise. The title of this example comes from the fact that the density function goes to ∞ as $y \to K$ so the lawn gets very soggy at the edge of the sprinkler's range. This is due to the fact that $s'(K) = \infty$, which in turn is caused by $r'(\pi/4) = 0$.

Our next example is a warm-up for (3.2).

Example 3.4. Suppose X has an exponential distribution with parameter 3. That is, X has density function $3e^{-3x}$ for $x \geq 0$. Let $Y = 1 - e^{-3X}$. Here, $r(x) = 1 - e^{-3x}$ is increasing on $(0, \infty)$, $\alpha = r(0) = 0$, and $\beta = r(\infty) = 1$. To find the inverse function we set $y = 1 - e^{-3x}$ and solve to get $s(y) = (-1/3)\ln(1-y)$. Differentiating, we have $s'(y) = -(-1/3)/(1 - y)$. So plugging into (3.1), the density function of Y is

$$f(s(y))s'(y) = 3e^{\ln(1-y)} \cdot \frac{1/3}{(1 - y)} = 1$$

for $0 < y < 1$. That is, Y is uniform on $(0, 1)$. There is nothing special about $\lambda = 3$ here. The next result shows that there is nothing very special about the exponential distribution.

(3.2) Suppose X has a continuous distribution. Then $Y = F(X)$ is uniform on $(0, 1)$.

PROOF: Even though F may not be strictly increasing, we can define an inverse of F by

$$F^{-1}(y) = \min\{x : F(x) \geq y\}$$

Using this definition of F^{-1}, we have

$$P(Y \le y) = P(X \le F^{-1}(y)) = F(F^{-1}(y)) = y$$

the last equality holding since F is continuous.   □

Reversing the ideas in the last proof, we get a result that is useful to construct random variables with a specified distribution.

(3.3) Suppose U has a uniform distribution on (0,1). Then $Y = F^{-1}(U)$ has distribution function F.

PROOF: The definition of F^{-1} was chosen so that if $0 < x < 1$ then

$$F^{-1}(y) \le x \text{ if and only if } F(x) \le y$$

and this holds for any distribution function F. Taking $y = U$, it follows that

$$P(F^{-1}(U) \le x) = P(U \le F(x)) = F(x)$$

since $P(U \le u) = u$.   □

For a concrete example, suppose we want to construct an exponential distribution with parameter λ. Setting $1 - e^{-\lambda x} = y$ and solving gives $-\ln(1-y)/\lambda = x$. So if U is uniform on $(0, 1)$ then $-\ln(1 - U)/\lambda$ has the desired exponential distribution.

EXERCISES

3.1. Suppose X has density function $f(x)$ for $a \le x \le b$ and $Y = cX + d$ where $c > 0$. Find the density function of Y.

3.2. Show that if $X = $ exponential(1) then $Y = X/\lambda$ is exponential(λ).

3.3. Suppose X is uniform on $(0, 1)$. Find the density function of $Y = X^n$.

3.4. Suppose X has density x^{-2} for $x \ge 1$ and $Y = X^{-2}$. Find the density function of Y.

3.5. Suppose X has an exponential distribution with parameter λ and $Y = X^{1/\alpha}$. Find the density function of Y. This is the **Weibull distribution**.

3.6. Suppose X has an exponential distribution with parameter 1 and $Y = \ln(X)$. Find the distribution function of X. This is the **double exponential distribution**.

3.7. Suppose X has a normal distribution and $Y = e^X$. Find the density function of Y. This is the **lognormal distribution**.

3.8. A drunk standing one foot from a wall shines a flashlight at a random angle that is uniformly distributed between $-\pi/2$ and $\pi/2$. Find the density function of the place where the light hits the wall. The answer is called the **Cauchy density**.

3.9. Suppose X is uniform on $(0, \pi/2)$ and $Y = \sin X$. Find the density function of Y. The answer is called the **arcsine law** because the distribution function contains the arcsine function.

3.10. Suppose X has density function $3x^{-4}$ for $x \geq 1$. (a) Find a function g so that $g(X)$ is uniform on $(0, 1)$. (b) Find a function h so that if U is uniform on $(0, 1)$, $h(U)$ has density function $3x^{-4}$ for $x \geq 1$.

3.11. Suppose X has density function $f(x)$ for $-1 \leq x \leq 1$, 0 otherwise. Find the density function of (a) $Y = |X|$, (b) $Z = X^2$.

3.12. Suppose X has density function $x/2$ for $0 < x < 2$, 0 otherwise. Find the density function of $Y = X(2 - X)$ by computing $P(Y \geq y)$ and then differentiating.

3.13. A weather channel has the local forecast on the hour and at 10, 25, 30, 45, and 55 minutes past. Suppose you wake up in the middle of the night and turn on the TV, and let X be the time you have to wait to see the local forecast, measured in hours. Find the density function of X.

3.14. Suppose r is differentiable and $\{x : r'(x) = 0\}$ is finite. Show that the density function of $Y = r(X)$ is given by

$$\sum_{x:r(x)=y} f(x)/|r'(x)|$$

3.15. Show that if $F_1(x) \leq F_2(x)$ are two distribution functions then by using the recipe in (3.3) we can define random variables X_1 and X_2 with these distributions so that $X_1 \geq X_2$.

3.4. Joint Distributions

In many situations we need to know the relationship between several random variables $X_1, \ldots, X_n$. We begin by considering examples of the case $n = 2$.

Example 4.1. If we are interested in determining how many American men are overweight and to what extent, then we need to look at the joint distribution of height and weight. A real chart would have a huge number of entries, so we will consider a fictional one in which height is measured to the nearest 0.5 foot and weight is measured to the nearest multiple of 25 pounds.

X	Y=100	125	150	175
5	.08	.08	.06	0
5.5	.08	.16	.16	.08
6	0	.08	.10	.12

When X and Y have discrete distributions, there is not much to say about the joint distribution. One makes a table that gives the probabilities of the various combinations of X and Y values. Moving on to the more complicated continuous case, two random variables are said to have **joint density function** f if

(4.1)
$$P((X,Y) \in A) = \iint_A f(x,y)\,dx\,dy$$

where

(4.2)
$$f(x,y) \ge 0 \quad \text{and} \quad \iint f(x,y)\,dx\,dy = 1$$

In words, we find the probability that (X,Y) lies in A by integrating f over A. As we will see a number of times below, it is useful to think of $f(x,y)$ as $P(X = x, Y = y)$ even though the last event has probability 0. As in Section 3.2, the precise interpretation of $f(x,y)$ is

$$P(x \le X \le x + \Delta x,\, y \le Y \le y + \Delta y) = \int_x^{x+\Delta x} \int_y^{y+\Delta y} f(u,v)\,dv\,du$$
$$\approx f(x,y)\Delta x \Delta y$$

when Δx and Δy are small, so $f(x,y)$ indicates how likely it is for (X,Y) to be near (x,y).

For a concrete example of a joint density function, consider

Example 4.2.
$$f(x,y) = \begin{cases} e^{-y} & 0 < x < y < \infty \\ 0 & \text{otherwise} \end{cases}$$

The story behind this example will be told later. To check that f is a density function, we observe that

$$\int_0^\infty \int_0^y e^{-y}\,dx\,dy = \int_0^\infty y e^{-y}\,dy$$

and integrating by parts with $g(y) = y$, $h'(y) = e^{-y}$ (so $g'(y) = 1$, $h(y) = -e^{-y}$).

$$\int_0^\infty ye^{-y}\,dy = -ye^{-y}\big|_0^\infty + \int_0^\infty e^{-y}\,dy = 0 + (-e^{-y})\big|_0^\infty = 1$$

To illustrate the use of (4.1) we will now compute $P(X \le 1)$, which can be written as $P((X,Y) \in A)$ where $A = \{(x,y) : x \le 1\}$. The formula in (4.1) tells us that we find $P((X,Y) \in A)$ by integrating the joint density over A. However, the joint density is only positive on $B = \{(x,y) : 0 < x < y < \infty\}$ so we only need to integrate over $A \cap B = \{(x,y) : 0 < x \le 1, x < y\}$, and doing this we find

$$P(X \le 1) = \int_0^1 \int_x^\infty e^{-y}\,dy\,dx$$

To evaluate the double integral we begin by observing that

$$\int_x^\infty e^{-y}\,dy = (-e^{-y})\big|_x^\infty = 0 - (-e^{-x}) = e^{-x}$$

so $P(X < 1) = \int_0^1 e^{-x}\,dx = (-e^{-x})\big|_0^1 = 1 - e^{-1}$.

Example 4.3. Pick a point at random from the ball $B = \{(x,y) : x^2 + y^2 \le 1\}$. By "at random from B" we mean that a choice outside of B is impossible and that all the points in B should be equally likely. In terms of the joint density this means that $f(x,y) = 0$ when $(x,y) \notin B$ and there is a constant $c > 0$ so that $f(x,y) = c$ when $(x,y) \in B$. Our $f(x,y) \ge 0$. To make the integral of f equal to 1, we have to choose c appropriately. Now,

$$\iint f(x,y)\,dx\,dy = \iint_B c\,dx\,dy = c \text{ (area of } B) = c\pi$$

So we choose $c = 1/\pi$ to make the integral 1 and define

$$f(x,y) = \begin{cases} 1/\pi & x^2 + y^2 \le 1 \\ 0 & \text{otherwise} \end{cases}$$

The arguments that led to the last conclusion generalize easily to show that if we pick a point "at random" from a set S with area a then

(4.3)
$$f(x,y) = \begin{cases} 1/a & (x,y) \in S \\ 0 & \text{otherwise} \end{cases}$$

Example 4.4. Buffon's needle. A floor consists of boards of width 1. If we drop a needle of length $L \le 1$ on the floor, what is the probability it will touch one of the cracks (i.e., the small spaces between the boards)? To make the question simpler to answer, we assume that the needle and the cracks have width zero.

Let X be the distance from the center of the needle to the nearest crack and Θ be the angle $\in [0, \pi)$ that the top half of the needle makes with the crack. (We make this choice to have $\sin \Theta > 0$.) We assume that all the ways the needle can land are equally likely, that is, the joint distribution of (X, Θ) is

$$f(x, \theta) = \begin{cases} 2/\pi & \text{if } x \in [0, 1/2), \ \theta \in [0, \pi) \\ 0 & \text{otherwise} \end{cases}$$

The formula for the joint density follows from (4.3). We are picking a point "at random" from a set S with area $\pi/2$, so the joint density is $2/\pi$ on S.

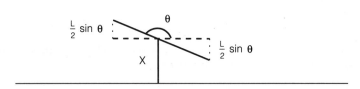

By drawing a picture (like the one above), one sees that the needle touches the crack if and only if $(L/2) \sin \Theta \geq X$. (4.1) tells us that the probability of this event is obtained by integrating the joint density over

$$A = \{(x, \theta) \in [0, 1/2) \times [0, \pi) : x \leq (L/2) \sin \theta\}$$

so the probability we seek is

$$\iint_A f(x, \theta) \, dx \, d\theta = \int_0^\pi \int_0^{(L/2) \sin \theta} \frac{2}{\pi} \, dx \, d\theta$$

$$= \frac{2}{\pi} \int_0^\pi \frac{L}{2} \sin \theta \, d\theta = \frac{L}{\pi}(-\cos \theta)\Big|_0^\pi = 2L/\pi$$

Buffon wanted to use this as a method of estimating π. Taking $L = 1/2$ and performing the experiment 10,000 times on a computer, we found that 1 over the fraction of times the needle hit the crack was 3.2310, 3.1368, and 3.0893 in the three times we tried this. We will see in Chapter 5 that these numbers are typical outcomes and that to compute π to 4 decimal places would require about 10^8 (or 100 million) tosses.

EXERCISE 4.1. Find an expression for the probability that a needle of length $L > 1$ touches the crack.

EXERCISE 4.2. **Buffon's quarter.** A tile floor consists of squares of side 1. Suppose we drop a quarter of diameter $r \leq 1$ on the floor. What is the probability it will not touch one of the cracks between the tiles?

REMARK. Before leaving the subject of joint densities, we would like to make one remark that will be useful later. If X and Y have joint density $f(x, y)$ then $P(X = Y) = 0$. To see this, we observe that $\iint_A f(x, y)\, dx\, dy$ is the volume of the region over A underneath the graph of f, but this volume is 0 if A is the line $x = y$.

The joint distribution of two random variables is occasionally described by giving the **joint distribution function**:

$$F(x, y) = P(X \leq x, Y \leq y)$$

The next example illustrates this notion but also shows that in this situation the density function is easier to write down.

Example 4.5. Suppose (X, Y) is uniformly distributed over the square $\{(x, y) : 0 < x < 1, 0 < y < 1\}$. That is,

$$f(x, y) = \begin{cases} 1 & 0 < x < 1, \ 0 < y < 1 \\ 0 & \text{otherwise} \end{cases}$$

Here, we are picking a point "at random" from a set with area 1, so the formula follows from (4.3). By patiently considering the possible cases, one finds that

$$F(x, y) = \begin{cases} 0 & \text{if } x < 0 \text{ or } y < 0 \\ xy & \text{if } 0 \leq x \leq 1 \text{ and } 0 \leq y \leq 1 \\ x & \text{if } 0 \leq x \leq 1 \text{ and } y > 1 \\ y & \text{if } x > 1 \text{ and } 0 \leq y \leq 1 \\ 1 & \text{if } x > 1 \text{ and } y > 1 \end{cases}$$

The first case should be clear: If $x < 0$ or $y < 0$ then $\{X \leq x, Y \leq y\}$ is impossible since X and Y always lie between 0 and 1. For the second case we note that when $0 \leq x \leq 1$ and $0 \leq y \leq 1$,

$$P(X \leq x, Y \leq y) = \int_0^x \int_0^y 1\, dv\, du = xy$$

In the third case, since values of $Y > 1$ are impossible,

$$P(X \leq x, Y \leq y) = P(X \leq x, Y \leq 1) = x$$

by the formula for the second case. The fourth case is similar to the third, and the fifth is trivial. X and Y are always smaller than 1 so if $x > 1$ and $y > 1$ then $\{X \leq x, Y \leq y\}$ has probability 1.

We will not use the joint distribution function in what follows. For completeness, however, we want to mention two of its important properties: (4.4) and (4.5). The first formula is the two-dimensional generalization of $P(a < X \leq b) = F(b) - F(a)$.

(4.4) $P(a_1 < X \leq b_1, a_2 < Y \leq b_2)$
$$= F(b_1, b_2) - F(a_1, b_2) - F(b_1, a_2) + F(a_1, a_2)$$

PROOF: The reasoning we use here is much like that employed in studying the probabilities of unions in Section 1.6. By adding and subtracting the probabilities on the right, we end up with the desired area counted exactly once. Let

$$B = (-\infty, a_1] \times (a_2, b_2] \qquad A = (a_1, b_1] \times (a_2, b_2]$$
$$D = (-\infty, a_1] \times (-\infty, a_2] \qquad C = (a_1, b_1] \times (-\infty, a_2]$$

See above. Using A as shorthand for $P((X,Y) \in A)$, etc.,

$$F(b_1, b_2) = A + B + C + D$$
$$-F(a_1, b_2) = -B - D$$
$$-F(b_1, a_2) = -C - D$$
$$F(a_1, a_2) = D$$

Adding the last four equations gives the one in (4.4).   □

The next formula tells us how to recover the joint density function from the joint distribution function. To motivate the formula, we recall that in one dimension $F' = f$ since $F(x) = \int_{\infty}^{x} f(u)\, du$.

$$(4.5) \qquad\qquad \frac{\partial^2 F}{\partial x \partial y} = f$$

To explain why this formula is true, we note that

$$F(x, y) = \int_{-\infty}^{x} \int_{-\infty}^{y} f(u, v)\, dv\, du$$

and differentiating twice kills the two integrals. To check that (4.5) works in Example 4.5, $F(x, y) = xy$ when $0 < x < 1$ and $0 < y < 1$, so $\frac{\partial^2 F}{\partial x \partial y} = 1$ there and it is 0 otherwise.

The developments above generalize in a straightforward way to $n > 2$ random variables $X_1, \ldots, X_n$. f is a **joint density function** if $f(x_1, \ldots, x_n) \geq 0$ and

$$\int \cdots \int f(x_1, \ldots, x_n)\, dx_n \cdots dx_1 = 1$$

The joint distribution function is defined by

$$F(x_1, \ldots, x_n) = P(X_1 \leq x_1, \ldots, X_n \leq x_n)$$
$$= \int_{-\infty}^{x_1} \cdots \int_{-\infty}^{x_n} f(y_1, \ldots, y_n)\, dy_n \cdots dy_1$$

and differentiating the last equality n times gives

$$(4.6) \qquad\qquad \frac{\partial^n F}{\partial x_1 \cdots x_n} = f$$

EXERCISE 4.3. Prove the analogue of (4.4) for $n = 3$ variables:

$$P(a_i < X_i \leq b_i \text{ for } i = 1, 2, 3) = F(b_1, b_2, b_3)$$
$$- F(a_1, b_2, b_3) - F(b_1, a_2, b_3) - F(b_1, b_2, a_3)$$
$$+ F(a_1, a_2, b_3) + F(a_1, b_2, a_3) + F(b_1, a_2, a_3)$$
$$- F(a_1, a_2, a_3)$$

To make the generalization to n variables clear we note that we have evaluated F at the eight points (c_i, d_i, e_i) with $c, d, e \in \{a, b\}$ and the sign of each term is $(-1)^{\#(a)}$ where $\#(a)$ is the number of a's.

EXERCISES

4.4. Suppose we draw 2 balls out of an urn with 8 red, 6 blue, and 4 green balls. Let X be the number of red balls we get and Y the number of blue balls. Find the joint distribution of X and Y.

4.5. Suppose we roll two dice that have the numbers 1, 2, 3, and 4 on their four sides. Let X be the maximum of the two numbers that appear and Y be the sum. Find the joint distribution of X and Y.

4.6. Suppose we roll two ordinary six-sided dice, let X be the minimum of the two numbers that appear, and let Y be the maximum of the two numbers. Find the joint distribution of X and Y.

4.7. Suppose we roll one die repeatedly and let N_i be the number of the roll on which i first appears. Find the joint distribution of N_1 and N_6.

4.8. Suppose $P(X = x, Y = y) = c(x + y)$ for $x, y = 0, 1, 2, 3$. (a) What value of c will make this a probability function? (b) What is $P(X > Y)$?

4.9. Suppose X and Y have joint density $f(x, y) = c(x + y)$ for $0 < x, y < 1$. (a) What is c? (b) What is $P(X < 1/2)$?

4.10. Suppose X and Y have joint density $f(x, y) = 6xy^2$ for $0 < x, y < 1$. What is $P(X + Y < 1)$?

4.11. Suppose X and Y have joint density $f(x, y) = 2$ for $0 < y < x < 1$. Find $P(X - Y > z)$.

4.12. We take a stick of length 1 and break it into 3 pieces. To be precise, we think of taking the unit interval and cutting at $X < Y$ where X and Y have joint density $f(x, y) = 2$ for $0 < x < y < 1$. What is the probability we can make a triangle with the three pieces? (We can do this if no piece is longer than 1/2.)

4.13. Suppose X and Y have joint density $f(x, y) = 1$ for $0 < x, y < 1$. Find $P(XY \le z)$.

4.14. X, Y, and Z have a uniform density on the unit cube. That is, their joint density is 1 when $0 < x, y, z < 1$ and 0 otherwise. Find $P(X + Y + Z < 1)$.

4.15. Suppose X and Y have joint density $f(x, y) = e^{-(x+y)}$ for $x, y > 0$. Find the distribution function.

4.16. Suppose X is uniform on $(0,1)$ and $Y = X$. Find the joint distribution function of X and Y.

4.17. A pair of random variables X and Y take values between 0 and 1 and have $P(X \le x, Y \le y) = x^3 y^2$ when $0 \le x, y \le 1$. Find the joint density function.

3.5. Marginal Distributions, Independence

The first question to be addressed in this section is: Given the joint distribution of (X, Y), how do we recover the distributions of X and Y? In the discrete case this is easy. The **marginal distributions** of X and Y are given by

(5.1)
$$P(X = x) = \sum_y P(X = x, Y = y)$$

$$P(Y = y) = \sum_x P(X = x, Y = y)$$

To explain the first formula in words, if $X = x$ then Y will take on some value y, so to find $P(X = x)$ we sum the probabilities of the disjoint events $\{X = x, Y = y\}$ over all the values of y.

Example 5.1. To illustrate these formulas we look at Example 4.1:

X	Y=0	1	2	3	
5	.08	.08	.06	0	.22
5.5	.08	.16	.16	.08	.48
6	0	.08	.10	.12	.30
	.16	.32	.32	.20	

In this case
$$P(X = 1) = .08 + .08 + .06 + 0 \ = .22$$
$$P(X = 2) = .08 + .16 + .16 + .08 = .48$$
$$P(X = 3) = 0 \ + .08 + .10 + .12 = .30$$

or in words, we find the distribution of X by summing the values in each row. Similarly, we sum the columns to find the distribution of Y:

$$P(Y = 0) = .08 + .08 + 0 \ = .16$$
$$P(Y = 1) = .08 + .16 + .08 = .32$$
$$P(Y = 2) = .06 + .16 + .10 = .32$$
$$P(Y = 3) = 0 \ + .08 + .12 = .20$$

You should recognize the distributions of X and Y as the numbers we wrote in the margins of the table. This is one explanation for the term "marginal distribution."

Formula (5.1) generalizes in a straightforward way to continuous distributions: we replace the sum by an integral and the probability functions by density functions. Suppose X and Y have joint density $f(x, y)$; then the **marginal densities** of X and Y are given by

(5.2)
$$f_X(x) = \int f(x, y)\, dy$$
$$f_Y(y) = \int f(x, y)\, dx$$

The verbal explanation of the first formula is similar to that of the discrete case: if $X = x$ then Y will take on some value y, so to find $P(X = x)$ we integrate the joint density $f(x, y)$ over all possible values of y. As advertised earlier, these formulas are easy to remember if you think of $f_X(x)$ as $P(X = x)$, $f_Y(y)$ as $P(Y = y)$, and $f(x, y)$ as $P(X = x, Y = y)$, for then they just look like (5.1) with an integral instead of a sum.

To illustrate the use of these formulas we look at Example 4.2.

Example 5.2.
$$f(x, y) = \begin{cases} e^{-y} & 0 < x < y < \infty \\ 0 & \text{otherwise} \end{cases}$$

In this case
$$f_X(x) = \int_x^\infty e^{-y}\, dy = (-e^{-y})\big|_x^\infty = e^{-x}$$

since (5.2) tells us to integrate $f(x, y)$ over all values of y but we only have $f > 0$ when $y > x$. Similarly,

$$f_Y(y) = \int_0^y e^{-y}\, dx = y e^{-y}$$

Two random variables are said to be **independent** if for any two sets A and B we have

(5.3)
$$P(X \in A, Y \in B) = P(X \in A) P(Y \in B)$$

In the discrete case (5.3) is equivalent to

(5.4)
$$P(X = x, Y = y) = P(X = x) P(Y = y)$$

for all x and y. To see this we note that taking $A = \{x\}$ and $B = \{y\}$ in (5.3) gives (5.4), while in the other direction if (5.4) holds,

$$P(X \in A, Y \in B) = \sum_{x \in A} \sum_{y \in B} P(X = x, Y = y)$$

$$= \sum_{x \in A} \sum_{y \in B} P(X = x)P(Y = y)$$

$$= \sum_{x \in A} P(X = x) \sum_{y \in B} P(Y = y) = P(X \in A)P(Y \in B)$$

To see what (5.4) says, we look at some examples

Example 5.3.

X	Y=1	2	
1	.3	0	.3
2	.4	.3	.7
	.7	.3	

In this case X and Y are not independent since

$$P(X = 1, Y = 2) = 0 < P(X = 1)P(Y = 2)$$

In words, if there is a 0 in the joint probability function of (X, Y) then X and Y are not independent.

Example 5.4.

X	Y=1	2	
1	.3	.1	.4
2	.4	.2	.6
	.7	.3	

In this case all of the entries are positive but X and Y are not independent since

$$P(X = 1, Y = 2) = 0.1 \neq 0.12 = P(X = 1)P(Y = 2)$$

If we want independent random variables with these marginal distributions then there is only one way to fill in the table:

Example 5.5.

X	Y=1	2	
1	.28	.12	.4
2	.42	.18	.6
	.7	.3	

In this case X and Y are independent since each entry in the table is the product of the marginal densities.

The next result is the continuous analogue of (5.4):

(5.5) Two random variables with joint density f are independent if and only if

$$f(x,y) = f_X(x)f_Y(y)$$

that is, if the joint density is the product of the marginal densities.

We will now consider three examples that parallel the ones used in the discrete case.

Example 5.6.

$$f(x,y) = \begin{cases} e^{-y} & 0 < x < y < \infty \\ 0 & \text{otherwise} \end{cases}$$

In this case

$$f(3,2) = 0 \neq f_X(3)f_Y(2) > 0$$

so (5.5) implies that X and Y are not independent. In general, if the set of values where $f > 0$ is not a rectangle then X and Y are not independent.

Example 5.7.

$$f(x,y) = \begin{cases} (1+x+y)/2 & 0 < x < 1 \text{ and } 0 < y < 1 \\ 0 & \text{otherwise} \end{cases}$$

In this case the set where $f > 0$ is a rectangle, so the joint distribution passes the first test and we have to compute the marginal densities

$$f_X(x) = \int_0^1 (1+x+y)/2\,dy = \left(\frac{1+x}{2}\right)y + \frac{y^2}{4}\bigg|_0^1 = \frac{x}{2} + \frac{3}{4}$$

$$f_Y(y) = \frac{y}{2} + \frac{3}{4} \qquad \text{by symmetry}$$

These formulas are valid for $0 < x < 1$ and $0 < y < 1$ respectively. To check independence we have to see if

$$(\star) \qquad \frac{1+x+y}{2} = \left(\frac{x}{2} + \frac{3}{4}\right) \cdot \left(\frac{y}{2} + \frac{3}{4}\right)$$

Multiplying both sides by 4 and simplifying the right-hand side, we see this holds if and only if

$$2 + 2x + 2y = 4xy + 6x + 6y + 9$$

for all $0 < x < 1$ and $0 < y < 1$, which is ridiculous. A simpler way to see that $(\star)$ is wrong is simply to note that when $x = y = 0$ it says that $1/2 = 9/16$.

Example 5.8.

$$f(x,y) = \begin{cases} \frac{y^{-3/2}\cos x}{(e-1)\sqrt{2\pi}} e^{\sin x - (1/2y)} & 0 < x < \pi/2, \ y > 0 \\ 0 & \text{otherwise} \end{cases}$$

In this case it would not be very much fun to integrate to find the marginal densities, so we adopt another approach.

(5.6) If $f(x,y)$ can be written as $g(x)h(y)$ then there is a constant c so that $f_X(x) = cg(x)$ and $f_Y(y) = h(y)/c$. It follows that $f(x,y) = f_X(x)f_Y(y)$ and hence X and Y are independent.

In words, if we can write f as a product of a function of x and a function of y then these functions must be constant multiples of the marginal densities. (5.6) takes care of our example since

$$f(x,y) = \left(\cos x \, e^{\sin x}\right)\left(\frac{y^{-3/2}}{(e-1)\sqrt{2\pi}} e^{-(1/2y)}\right)$$

PROOF: We begin by observing

$$f_X(x) = \int f(x,y)\,dy = g(x)\int h(y)\,dy$$

$$f_Y(y) = \int f(x,y)\,dx = h(y)\int g(x)\,dx$$

$$1 = \int\int f(x,y)\,dx\,dy = \int g(x)\,dx\int h(y)\,dy$$

So if we let $c = \int h(y)\,dy$ then the last equation implies $\int g(x)\,dx = 1/c$, and the first two give us $f_X(x) = cg(x)$ and $f_Y(y) = h(y)/c$. □

The concepts in this section generalize easily to $n > 2$ random variables. In the discrete case, we find the marginal distribution of one variable by summing over all the possible values of the other variables. For example,

$$P(X_1 = x_1) = \sum_{y_2,y_3,y_4} P(X_1 = x_1, X_2 = y_2, X_3 = y_3, X_4 = y_4)$$

or if we want to find the joint density of (X_1, X_3),

$$P(X_1 = x_1, X_3 = x_3) = \sum_{y_2,y_4} P(X_1 = x_1, X_2 = y_2, X_3 = x_3, X_4 = y_4)$$

In the continuous case the recipe is the same but we integrate out the unwanted variables:

$$f_{X_1, X_3}(x_1, x_3) = \iint f(x_1, y_2, x_3, y_4)\, dy_2\, dy_4$$

Several random variables $X_1, \ldots, X_n$ are said to be **independent** if

$$P(X_1 \in A_1, \ldots, X_n \in A_n) = P(X_1 \in A_1) \cdots P(X_n \in A_n)$$

(5.4)–(5.6) extend in the obvious way to $n > 2$ variables.

(5.4') Discrete random variables $X_1, \ldots, X_n$ are independent if and only if

$$P(X_1 = x_1, \ldots, X_n = x_n) = P(X_1 = x_1) \cdots P(X_n = x_n)$$

Continuous random variables are independent if and only if the joint density

(5.5') $$f(x_1, \ldots, x_n) = f_{X_1}(x_1) \cdots f_{X_n}(x_n)$$

or if we can write

(5.6') $$f(x_1, \ldots, x_n) = g_1(x_1) \cdots g_n(x_n)$$

Turning to examples:

Example 5.9. Let U be uniform on $(0, 1)$ and let X_k be the kth digit in the decimal expansion of U. For any n, $X_1, \ldots, X_n$ are independent. To check (5.4') we note that

$$P(X_1 = 3, X_2 = 7, X_3 = 8, X_4 = 1) = P(U \in [0.3781, 0.3782)) = 10^{-4}$$
$$= P(X_1 = 3)P(X_2 = 7)P(X_3 = 8)P(X_4 = 1)$$

or more generally, if $k_1, \ldots, k_n \in \{0, 1, \ldots, 9\}$

$$P(X_1 = k_1, \ldots, X_n = k_n) = 10^{-n} = P(X_1 = k_1) \cdots P(X_n = k_n)$$

Example 5.10. This example gives a remarkable property of the Poisson distribution. Let $A_1, \ldots, A_k$ be disjoint events whose union $\cup_{i=1}^{k} A_i = \Omega$. Suppose we perform the experiment a random number of times N, where N has a Poisson distribution with mean λ, and let X_i be the number of times A_i occurs. If

$n = x_1 + \cdots + x_k$, then recalling the formula for the multinomial distribution (Example 1.9 in Chapter 2),

$$P(X_i = x_i \text{ for } 1 \le i \le k) = e^{-\lambda} \frac{\lambda^n}{n!} \,'\, \frac{n!}{x_1! \cdots x_k!} P(A_1)^{x_1} \cdots P(A_k)^{x_k}$$

$$= e^{-\lambda P(A_1)} \frac{(\lambda P(A_1))^{x_1}}{x_1!} \cdots e^{-\lambda P(A_k)} \frac{(\lambda P(A_k))^{x_k}}{x_k!}$$

since $\sum_{i=1}^{k} P(A_i) = 1$. In words, $X_1, \ldots, X_k$ are independent Poissons with parameters $\lambda P(A_i)$.

To see why this is surprising, consider the special case $k = 2$, i.e., $A_2 = A_1^c$. If we performed our experiment a fixed number of times then N_1 and N_2 would not be independent since $N_2 = n - N_1$. It is remarkable that when we perform our experiment a Poisson number of times, the number of successes tells us nothing about the number of failures. This result is not only surprising but also useful. For a concrete example, suppose that a Poisson number of cars arrive at a fast food restaurant each hour and let A_i be the event that the car has i passengers. Then the number of cars with i passengers that arrive are independent Poissons.

The next two results say that functions of independent random variables are independent.

(5.7) Suppose $X_1, \ldots, X_n$ are independent. Then $r_1(X_1), \ldots, r_n(X_n)$ are independent.

PROOF: If we let $r^{-1}(A) = \{y : r(y) \in A\}$ then

$$P(r_i(X_i) \in A_i \text{ for } i = 1, \ldots, n) = P(X_i \in r_i^{-1}(A_i) \text{ for } i = 1, \ldots, n)$$

$$= P(X_1 \in r_1^{-1}(A_1)) \cdots P(X_n \in r_n^{-1}(A_n))$$

$$= P(r_1(X_1) \in A_1) \cdots P(r_n(X_n) \in A_n)$$

by applying the definition of independence to the sets $r^{-1}(A_i)$. $\square$

Given (5.7), the reader should not be surprised at

(5.8) If $X_1, \ldots, X_{m+n}$ are independent then so are

$$f(X_1, \ldots, X_n) \text{ and } g(X_{n+1}, \ldots, X_{n+m})$$

SKETCH OF PROOF: It should be easy to believe that $(X_1, \ldots, X_n)$ and $(X_{n+1}, \ldots, X_{n+m})$ are independent in the sense that

$$P((X_1, \ldots, X_n) \in B_1, (X_{n+1}, \ldots, X_{n+m}) \in B_2)$$

$$= P((X_1, \ldots, X_n) \in B_1) P((X_{n+1}, \ldots, X_{n+m}) \in B_2)$$

for all sets $B_1 \subset \mathbf{R}^n$ and $B_2 \subset \mathbf{R}^m$. Once one accepts this fact (which we do not have the machinery to prove here), the rest is like (5.7):

$$P(f(X_1,\ldots,X_n) \in A_1, g(X_{n+1},\ldots,X_{n+m}) \in A_2)$$
$$= P((X_1,\ldots,X_n) \in f^{-1}(A_1), (X_{n+1},\ldots,X_{n+m}) \in g^{-1}(A_2))$$
$$= P((X_1,\ldots,X_n) \in f^{-1}(A_1))\, P((X_{n+1},\ldots,X_{n+m}) \in g^{-1}(A_2))$$
$$= P(f(X_1,\ldots,X_n) \in A_1)P(g(X_{n+1},\ldots,X_{n+m}) \in A_2) \qquad \square$$

EXERCISES

5.1. Suppose a point (X, Y) is chosen at random from the circle $x^2 + y^2 \leq 1$. Find the marginal density of X.

5.2. Suppose X and Y have joint density $f(x, y) = x + 2y^3$ when $0 < x < 1$ and $0 < y < 1$. Find the marginal densities of X and Y.

5.3. Suppose X and Y have joint density $f(x, y) = 6y$ when $x > 0$, $y > 0$, and $x + y < 1$. Find the marginal densities of X and Y.

5.4. Suppose X and Y have joint density $f(x, y) = 10x^2y$ when $0 < y < x < 1$. Find the marginal densities of X and Y.

5.5. Let (X, Y, Z) be a random point in the unit sphere. That is, their joint density is $3/4\pi$ when $x^2 + y^2 + z^2 \leq 1$, 0 otherwise. Find the marginal density of (a) (X, Y), (b) Z.

5.6. Given the joint distribution function $F_{X,Y}(x, y) = P(X \leq x, Y \leq y)$, how do you recover the marginal distribution $F_X(x) = P(X \leq x)$?

5.7. Suppose X and Y have joint density $f(x, y)$. Are X and Y independent if
(a) $f(x, y) = xe^{-x(1+y)}$ for $x, y \geq 0$?
(b) $f(x, y) = 6xy^2$ when $x, y \geq 0$ and $x + y \leq 1$?
(c) $f(x, y) = 2xy + x$ when $0 < x < 1$ and $0 < y < 1$?
(d) $f(x, y) = (x + y)^2 - (x - y)^2$ when $0 < x < 1$ and $0 < y < 1$?
In each case $f(x, y) = 0$ otherwise.

5.8. Two people agree to meet for a drink after work but they are impatient and each will only wait 15 minutes for the other person to show up. Suppose that they each arrive at independent random times uniformly distributed between 5 p.m. and 6 p.m. What is the probability they will meet?

5.9. Suppose X_1 and X_2 are independent and uniform on $(0,1)$. In Exercise 6.1 you will show that the joint density of $Y = X_1/X_2$ and $Z = X_1X_2$ is given by $f_{(Y,Z)} = 1/2y$ when $y > z > 0$ and $yz < 1$, 0 otherwise. Find the marginal densities of Y and Z.

5.10. Suppose X_1 and X_2 are independent and normal$(0,1)$. Find the distribution of $Y = (X_1^2 + X_2^2)^{1/2}$. This is the **Rayleigh distribution.**

5.11. Suppose X_1, X_2, X_3 are independent and normal$(0,1)$. Find the distribution of $Y = (X_1^2 + X_2^2 + X_3^2)^{1/2}$. This is the **Maxwell distribution.** It is used in physics for the speed of particles in a gas.

5.12. Suppose $X_1, \ldots, X_n$ are independent and have distribution function $F(x)$. Find the distribution functions of (a) $Y = \max\{X_1, \ldots, X_n\}$ and (b) $Z = \min\{X_1, \ldots, X_n\}$

5.13. Suppose $X_1, \ldots, X_n$ are independent exponential(λ). Show that

$$\min\{X_1, \ldots, X_n\} = \text{exponential}(n\lambda)$$

5.14. Suppose $X_1, X_2, \ldots$ are independent and have the same continuous distribution F. We say that a record occurs at time k if $X_k > \max_{j<k} X_j$. Show that the events $A_k =$ "a record occurs at time k" are independent and $P(A_k) = 1/k$.

5.15. Let $E_1, \ldots, E_n$ be events and let X_i be 1 if E_i occurs, 0 otherwise. These are called **indicator random variables,** since they indicate whether or not the ith event occurred. Show that the indicator random variables are independent if and only if the events E_i are.

*3.6. Functions of Several Random Variables

New examples of joint densities often arise from old ones after a change of variables. Let S be a subset of R^n so that $P((X_1, \ldots, X_n) \in S) = 1$ and suppose that the function r maps S 1-1 onto a set T. That is, $\{r(x) : x \in S\} = T$ and there is an inverse function s from T to S, so that $s(r(x)) = x$ for $x \in S$. Let s_i be the ith component of the inverse function, let $D_{ij} = \partial s_i/\partial y_j$ be the matrix of partial derivatives, and let $J = \det D$ be its determinant. (J is for Jacobian.) By using the change of variables formula from multivariate calculus one can prove that the density function of $Y = r(X)$ is given by

$$(6.1) \qquad f_Y(y) = \begin{cases} f_X(s(y))|J| & \text{for } y \in T \\ 0 & \text{otherwise} \end{cases}$$

We will not prove this, but the reader should note that (6.1) reduces to (3.1) in the case $n = 1$.

Example 6.1. Suppose X_1 and X_2 have joint density $f(x_1, x_2) = g(x_1)h(x_2)$. That is, X_1 and X_2 are independent. Let $r(x_1, x_2) = (x_1, x_1 + x_2)$. To find the inverse we set $y_1 = x_1$, $y_2 = x_1 + x_2$ and solve:

$$x_1 = y_1, \qquad x_2 = y_2 - x_1 = y_2 - y_1$$

So $s_1(y) = y_1$, $s_2(y) = y_2 - y_1$. Calculating partial derivatives, we find

$$D_{11} = \partial s_1/\partial y_1 = 1 \qquad D_{12} = \partial s_1/\partial y_2 = 0$$
$$D_{21} = \partial s_2/\partial y_1 = -1 \qquad D_{22} = \partial s_2/\partial y_2 = 1$$

Hence the determinant $J = D_{11}D_{22} - D_{12}D_{21} = 1$ and using (6.1),

(6.2) $$f_{(Y_1,Y_2)}(y_1, y_2) = g(y_1)h(y_2 - y_1)$$

When $g(x) = h(x) = e^{-x}$ for $x \geq 0$, $g(y_1)h(y_2 - y_1) = e^{-y_1}e^{-(y_2-y_1)} = e^{-y_2}$ so

$$f_{Y_1,Y_2}(y_1, y_2) = \begin{cases} e^{-y_2} & \text{for } 0 \leq y_1 \leq y_2 \\ 0 & \text{otherwise} \end{cases}$$

The conditions on the first line come from observing that $s_1(y) \geq 0$ and $s_2(y) \geq 0$ if and only if $y_1 \geq 0$ and $y_2 \geq y_1$. Notice that after a change of notation the density here is the one in Example 4.2. This gives the story we promised earlier: If X_1 and X_2 are independent exponentials with parameter 1, then $f(y_1, y_2) = e^{-y_2}$ for $0 < y_1 < y_2 < \infty$ gives the joint density of $Y_1 = X_1$ and $Y_2 = X_1 + X_2$.

One of the reasons for computing a joint density is to compute a marginal density. Integrating out y_1 in (6.2), we have

$$f_{Y_2}(y_2) = \int f_{(Y_1,Y_2)}(y_1, y_2)\, dy_1 = \int f_{X_1}(y_1) f_{X_2}(y_2 - y_1)\, dy_1$$

Changing variables $y = y_1$, $z = y_2$ and writing $X_1 + X_2$ instead of Y_2, we have

(6.3) $$f_{X_1+X_2}(z) = \int f_{X_1}(y) f_{X_2}(z - y)\, dy$$

a formula that will be useful in the next section.

For our next example we will consider the quotient of two independent random variables. The formula we will derive is useful in computing some important distributions in statistics.

Example 6.2. Suppose X_1 and X_2 are independent with density functions f_1 and f_2. Compute the joint density of X_1 and X_2/X_1 and then use this to find the density of X_2/X_1.

Let $r(x_1, x_2) = (x_1, x_2/x_1)$. To find the inverse we set $y_1 = x_1$, $y_2 = x_2/x_1$ and solve to find

$$x_1 = y_1, \qquad x_2 = x_1 y_2 = y_1 y_2$$

so $s_1(y) = y_1$, $s_2(y) = y_1y_2$. Calculating partial derivatives, we find

$$D_{11} = \partial s_1/\partial y_1 = 1 \qquad D_{12} = \partial s_1/\partial y_2 = 0$$
$$D_{21} = \partial s_2/\partial y_1 = y_2 \qquad D_{22} = \partial s_2/\partial y_2 = y_1$$

Hence the determinant $J = D_{11}D_{22} - D_{12}D_{21} = y_1$ and using (6.1),

$$f_{(Y_1,Y_2)}(y_1, y_2) = |y_1| f_1(y_1) f_2(y_1 y_2)$$

Integrating out y_1 in the last equation, we have

$$f_{Y_2}(y_2) = \int f_{(Y_1,Y_2)}(y_1, y_2)\, dy_1 = \int |y_1| f_1(y_1) f_2(y_1 y_2)\, dy_1$$

Changing variables $y = y_1$, $z = y_2$ and writing X_2/X_1 instead of Y_2, we have

(6.4)
$$f_{X_2/X_1}(z) = \int |y| f_1(y) f_2(yz)\, dy$$

Our final example is a very simple situation in which the function is not one-to-one.

Example 6.3. Order statistics. Suppose $X_1, \ldots, X_n$ are independent and have density function f. Let $X^{(1)}$ be the smallest of the X_j, $X^{(2)}$ be the second smallest, and so on until $X^{(n)}$ is the largest. (By a simple extension of the remark after Example 4.4, all the X's are different with probability 1.) $X^{(1)}, \ldots, X^{(n)}$ are called **order statistics.** Their joint density is given by

(6.5)
$$f(x^1, \ldots, x^n) = \begin{cases} n! f(x^1) \ldots f(x^n) & \text{if } x^1 < x^2 \ldots < x^n \\ 0 & \text{otherwise} \end{cases}$$

To see this, note that the the joint density of $X_1, \ldots, X_n$ is $f(x_1) \cdots f(x_n)$ and the function that takes a vector $(x_1, \ldots, x_n)$ of distinct numbers and rearranges its components in increasing order maps $n!$ points that have the same probability density into one. Taking $f(x) = e^{-x}$, and noticing that

$$x^n + \cdots + x^1 = (x^n - x^{n-1}) + 2(x^{n-1} - x^{n-2}) + 3(x^{n-2} - x^{n-3})$$
$$+ \cdots + (n-1)(x^2 - x^1) + nx^1$$

(6.5) can be written in this case as

(6.6)
$$e^{-(x^n - x^{n-1})} 2e^{-2(x^{n-1} - x^{n-2})} \cdots (n-1)e^{-(n-1)(x^2 - x^1)} ne^{-nx^1}$$

and it follows that the differences between the order statistics are independent exponentials.

To explain this miracle, we note that Exercise 5.13 shows that the minimum of n independent exponential(λ) random variables, $X^{(1)}$, is exponential($n\lambda$). The lack of memory property of the exponential discussed in Example 2.7 implies that if we condition on $X^{(1)} = X_j = x$ then the $X_i - x$ for $i \neq j$ are independent exponential(λ). So using Exercise 5.13 again, it follows that $X^{(2)} - X^{(1)}$ is exponential($(n-1)\lambda$). Similar reasoning applies to $X^{(3)} - X^{(2)}, \ldots, X^{(n)} - X^{(n-1)}$. The fact that the differences are independent also follows from the lack of memory property.

EXERCISES

6.1. Suppose X_1 and X_2 have joint density

$$f(x_1, x_2) = \begin{cases} 1 & \text{for } 0 < x_1, x_2 < 1 \\ 0 & \text{otherwise} \end{cases}$$

Find the joint density of $Y_1 = X_1/X_2$ and $Y_2 = X_1 X_2$.

6.2. Suppose X and Y are independent standard normals. Find the density of Y/X. This is the Cauchy density.

6.3. Suppose X_1 and X_2 have joint density

$$f(x_1, x_2) = \frac{\lambda^{m+n+2} x_1^m x_2^n}{m!\, n!} e^{-\lambda(x_1 + x_2)}$$

Find the joint density of $Y_1 = X_1 + X_2$ and $Y_2 = X_1/(X_1 + X_2)$ and show that Y_1 and Y_2 are independent.

6.4. Suppose X_1 and X_2 have joint density

$$f(x_1, x_2) = g\left(\sqrt{x_1^2 + x_2^2}\right)$$

Examples are (X_1, X_2) uniform on $\{(x_1, x_2) : x_1^2 + x_2^2 \leq 1\}$ or X_1 and X_2 independent and having the standard normal distribution. Find the joint density of $R = \sqrt{X_1^2 + X_2^2}$ and $\Theta = \tan^{-1}(y/x)$ and show R and Θ are independent.

6.5. Suppose X_1 and X_2 have joint density

$$f(x_1, x_2) = g\left(\sqrt{x_1^2 + x_2^2}\right)$$

Show that $Y_1 = \cos\theta X_1 + \sin\theta X_2$ and $Y_2 = -\sin\theta X_1 + \cos\theta X_2$ have the same joint distribution. In words, this distribution is invariant under rotation.

6.6. Suppose U_1 and U_2 have joint density $f(x,y) = 1$ for $0 < x, y < 1$. Then $X_1 = \sqrt{-2\log U_1}\,\cos(2\pi U_2)$ and $X_2 = \sqrt{-2\log U_1}\,\cos(2\pi U_2)$ have joint density.

$$f(x_1, x_2) = (2\pi)^{-1} e^{-(x_1^2 + x_2^2)/2}$$

That is, X_1 and X_2 are independent normals. Since there is no closed form expression for the normal distribution function, this provides an attractive alternative to (3.3) for generating random variables with normal distributions.

6.7. Suppose $X_1, \ldots, X_n$ are independent and have density function $f(x)$. Let $Y = \min\{X_1, \ldots, X_n\}$ and $Z = \max\{X_1, \ldots, X_n\}$. Compute $P(Y \geq y, Z \leq z)$ and differentiate to find the joint density of Y and Z.

6.8. Suppose $X_1, \ldots, X_n$ are independent and uniform on $(0,1)$. Let $X^{(k)}$ be the kth largest of the X_j. By computing the appropriate marginal of the joint distribution in (6.5), conclude that the density of $X^{(k)}$ is given by

$$n\binom{n-1}{k-1} x^{k-1}(1-x)^{n-k}$$

This formula is easy to understand: There are n values of j that we can pick to be $= x$ (which occurs with a probability density of 1). The rest of the formula then gives the probability that exactly $k-1$ of the remaining $n-1$ variables are smaller than x.

6.9. Generalize the result in the last problem to show that if $X_1, \ldots, X_n$ are independent and have density f then the density of $X^{(k)}$ is given by

$$nf(x)\binom{n-1}{k-1} F(x)^{k-1}(1-F(x))^{n-k}$$

6.10. Find the joint density of $X^{(j)}$ and $X^{(k)}$ for $j < k$ under the assumptions of the last problem.

6.11. Suppose $X_1, \ldots, X_5$ are independent and have a common distribution that is continuous. Show that $P(X_3 < X_5 < X_1 < X_4 < X_2) = 1/5!$

6.12. A machine has 5 components and needs at least 3 working components to function. Suppose that their lifetimes are independent exponential(1). Find the density function for the time to failure T.

3.7. Sums of Independent Random Variables

In this section we will compute the distribution of $X + Y$ when X and Y are independent. In the discrete case this is easy:

(7.1)
$$P(X + Y = z) = \sum_x P(X = x, Y = z - x)$$
$$= \sum_x P(X = x)P(Y = z - x)$$

To see the first equality, note that if the sum is z then X must take on some value x and Y must be $z - x$. The first equality is valid for any random variables. The second holds since we have supposed X and Y are independent.

Example 7.1. If $X = \text{binomial}(n, p)$ and $Y = \text{binomial}(m, p)$ are independent then $X + Y = \text{binomial}(n + m, p)$.

The easiest way to see the conclusion is to note that if X is the number of successes in the first n trials and Y is the number of successes in the next m trials, then $X + Y$ is the number of successes in $n + m$ trials.

To get the conclusion by computation we use (7.1), note that $P(X = j) = 0$ when $j < 0$, $P(Y = k - j) = 0$ when $j > k$, and plug in the definition of the binomial distribution to get

$$P(X + Y = k) = \sum_{j=0}^{k} P(X = j)P(Y = k - j)$$

$$= \sum_{j=0}^{k} \binom{n}{j} p^j (1 - p)^{n-j} \binom{m}{k - j} p^{k-j} (1 - p)^{m-(k-j)}$$

$$= p^k (1 - p)^{n+m-k} \sum_{j=0}^{k} \binom{n}{j} \binom{m}{k - j}$$

$$= p^k (1 - p)^{n+m-k} \binom{n + m}{k}$$

To see the last equality, note that we can pick k students out of a class of n boys and m girls in $C_{n+m,k}$ ways but this can be done by first deciding on the number j of boys to be chosen and then picking j of the n boys (which can be done in $C_{n,j}$ ways) and $k - j$ of the m girls (which can be done in $C_{m,k-j}$ ways). The multiplication rule implies that for fixed j the number of ways the j boys and $k - j$ girls can be selected is $C_{n,j}C_{m,k-j}$, so summing from $j = 0$ to

k gives

$$\sum_{j=0}^{k} \binom{n}{j}\binom{m}{k-j} = \binom{n+m}{k}$$

Example 7.2. If $X = \text{Poisson}(\lambda)$ and $Y = \text{Poisson}(\mu)$ are independent then $X + Y = \text{Poisson}(\lambda + \mu)$.

Again we use (7.1), note that $P(X = j) = 0$ when $j < 0$, $P(Y = k - j) = 0$ when $j > k$, and plug in the definition of the Poisson distribution to get

$$P(X + Y = k) = \sum_{j=0}^{k} P(X = j)P(Y = k - j)$$

$$= \sum_{j=0}^{k} e^{-\lambda}\frac{\lambda^j}{j!} e^{-\mu}\frac{\mu^{k-j}}{(k-j)!}$$

$$= e^{-(\lambda+\mu)}\frac{1}{k!}\sum_{j=0}^{k} \binom{k}{j}\lambda^j\mu^{k-j}$$

$$= e^{-(\lambda+\mu)}\frac{(\lambda+\mu)^k}{k!}$$

where the last equality follows from the binomial theorem ((3.5) in Chapter 1).

EXERCISE 7.1. Suppose $X = \text{Poisson}(\lambda)$ and $Y = \text{Poisson}(\mu)$ with $\lambda < \mu$. Show that for each k, $P(X \le k) \ge P(Y \le k)$.

Formula (7.1) generalizes in the usual way to continuous distributions: we replace the probabilities by density functions and the sum by an integral.

$$(7.2) \qquad f_{X+Y}(z) = \int f_X(x)f_Y(z-x)\,dx$$

This result was derived in Section 3.6; see (6.3). The next exercise gives another proof.

EXERCISE 7.2. Prove (7.2) by noticing that

$$P(X + Y \le z) = \int\int_{-\infty}^{z-x} f_X(x)f_Y(y)\,dy\,dx$$

$$= \int f_X(x)F_Y(z-x)\,dx$$

and then differentiating with respect to z.

For a concrete example of the use of (7.2), consider

Example 7.3. If $X = \text{uniform}(0,1)$ and $Y = \text{uniform}(0,1)$ then $X + Y$ has the **triangular density**

$$f_{X+Y}(z) = \begin{cases} z & 0 \le z \le 1 \\ 2 - z & 1 \le z \le 2 \\ 0 & \text{otherwise} \end{cases}$$

The integrand in (7.2) is 1 when $0 < x < 1$ and $0 < z - x < 1$. The second set of inequalities can be written as $z - 1 < x < z$. When $z \le 1$ the two inequalities combine to $0 < x < z$ so

$$f_{X+Y}(z) = \int_0^z 1 \, dx = z$$

When $1 \le z \le 2$ the two inequalities combine to $z - 1 < x < 1$ so

$$f_{X+Y}(z) = \int_{z-1}^1 1 \, dx = 2 - z$$

Since $0 < X + Y < 2$ the density function f_{X+Y} must be 0 otherwise.

Our next example gives a special and important property of the normal distribution.

Example 7.4. If $X_1 = \text{normal}(\mu, a)$ and $X_2 = \text{normal}(\nu, b)$ then $X_1 + X_2 = \text{normal}(\mu + \nu, a + b)$.

Suppose $Y_1 = \text{normal}(0, a)$ and $Y_2 = \text{normal}(0, b)$. Then (7.2) implies

$$f_{Y_1+Y_2}(z) = \frac{1}{2\pi\sqrt{ab}} \int e^{-x^2/2a} e^{-(z-x)^2/2b} \, dx$$

Dropping the constant in front, the integral can be rewritten as

$$\int \exp\left(-\frac{bx^2 + ax^2 - 2axz + az^2}{2ab}\right) dx$$

$$= \int \exp\left(-\frac{a+b}{2ab}\left\{x^2 - \frac{2a}{a+b}xz + \frac{a}{a+b}z^2\right\}\right) dx$$

$$= \int \exp\left(-\frac{a+b}{2ab}\left\{\left(x - \frac{a}{a+b}z\right)^2 + \frac{ab}{(a+b)^2}z^2\right\}\right) dx$$

since $-\{a/(a+b)\}^2 + \{a/(a+b)\} = ab/(a+b)^2$. Factoring out the term that does not depend on x, the last integral

$$= \exp\left(-\frac{z^2}{2(a+b)}\right) \int \exp\left(-\frac{a+b}{2ab}\left(x - \frac{a}{a+b}z\right)^2\right) dx$$

$$= \exp\left(-\frac{z^2}{2(a+b)}\right) \sqrt{2\pi ab/(a+b)}$$

since the last integral is the normal density with parameters $\mu = az/(a+b)$ and $\sigma^2 = ab/(a+b)$ without its proper normalizing constant. Reintroducing the constant we dropped at the beginning,

$$f_{Y_1+Y_2}(z) = \frac{1}{2\pi\sqrt{ab}} \sqrt{2\pi ab/(a+b)} \exp\left(-\frac{z^2}{2(a+b)}\right)$$

and we have proved the result when $\mu = \nu = 0$. To get the result in general, let $X_1 = \mu + Y_1$, $X_2 = \nu + Y_2$ and notice that

$$X_1 + X_2 = (\mu + \nu) + Y_1 + Y_2 = \text{normal}(\mu + \nu, a + b) \qquad \square$$

Example 7.5. Let $t_1, \ldots, t_n$ be random variables that are independent and have an exponential distribution with parameter λ. Then $T_n = t_1 + \cdots + t_n$ has the **gamma**(n, λ) density function

(7.3) $$f(x) = \frac{\lambda^n x^{n-1}}{(n-1)!} e^{-\lambda x}$$

PROOF: We will prove the result by induction. When $n = 1$, $T_1 = t_1$ and $f(x)$ is just the exponential density and so the result is true. Suppose now that it is true for n, and let $X = T_n$ and $Y = t_{n+1}$. Using (7.2) and the result for n variables, we have

$$f_{T_n+t_{n+1}}(z) = \int_0^z f_{T_n}(x) f_{t_{n+1}}(z-x) \, dx$$

$$= \int_0^z \frac{\lambda^n x^{n-1}}{(n-1)!} e^{-\lambda x} \cdot \lambda e^{-\lambda(z-x)} \, dx$$

$$= e^{-\lambda z} \frac{\lambda^{n+1}}{(n-1)!} \int_0^z x^{n-1} \, dx$$

$$= e^{-\lambda z} \frac{\lambda^{n+1}}{(n-1)!} \cdot \frac{z^n}{n} = e^{-\lambda z} \frac{\lambda^{n+1} z^n}{n!}$$

which is (7.3) with n replaced by $n + 1$, and x replaced by z.   □

EXERCISE 7.3. Suppose $X = \text{gamma}(n, 1)$. Show that $X/\lambda = \text{gamma}(n, \lambda)$.

EXERCISE 7.4. Suppose $X = \text{gamma}(m, \lambda)$ and $Y = \text{gamma}(n, \lambda)$ are independent. Then $X + Y = \text{gamma}(m + n, \lambda)$. As in Example 7.1, you can think or you can compute.

EXERCISES

7.5. Suppose X and Y are independent and take the values 1, 2, 3, and 4 with probabilities $0.1, 0.2, 0.3, 0.4$. Find the probability function for the sum $X + Y$.

7.6. Suppose X and Y are independent and have a geometric distribution with parameter p. Find the distribution of $X + Y$.

7.7. Suppose $X_1, \ldots, X_n$ are independent and have a geometric distribution with parameter p. $T = X_1 + \cdots + T_n$ is the amount of time we have to wait for n successes when each trial is independent and results in success with probability p. Use this interpretation to show

$$P(T = m) = \binom{m - 1}{n - 1} p^n (1 - p)^{m - n}$$

This is the negative binomial distribution introduced at the end of Section 2.1.

7.8. Suppose $X = \text{uniform on } (0,1)$ and $Y = \text{uniform on } (0,2)$ are independent. Find the density function of $X + Y$.

7.9. Suppose X_1, X_2, X_3 are independent and uniform on $(0,1)$. Find the density function of $S = X_1 + X_2 + X_3$. You can save yourself some work by noting that the density will be symmetric about $3/2$.

7.10. Suppose X and Y are independent, X is uniform on $(0, 1)$, and Y has density f and distribution function F. Show that $X + Y$ has density function $F(x) - F(x - 1)$.

7.11. Suppose X and Y are independent and have density function $f(x) = 2x$ for $0 < x < 1$. Find the density function of $X + Y$.

7.12. Suppose X and Y are independent and have exponential distributions with parameters $\lambda < \mu$. Find the density function of $X + Y$.

7.13. We have a flashlight that uses two batteries and we have a package of four new batteries. Suppose that each battery has an independent exponential(λ) lifetime. Compute the distribution of the number of hours T that we can use the flashlight.

Poisson process. As in Example 7.5, let $t_1, t_2, \ldots$ be independent exponential(λ) and $T_n = t_1 + \cdots + t_n$. We think of T_n as the arrival time of the nth customer at a bank or ice cream parlor. Let $N_t = \max\{n : T_n < t\}$ be the number of customers that have arrived by time t. To justify the name "Poisson process" we will now show that

(7.4) N_t has a Poisson(λt) distribution.

PROOF: To begin, we note that $P(N_t < n) = P(T_n > t)$. That is, fewer than n customers will arrive by time t if and only if the nth customer arrives after time t. To compute $P(T_n > t)$ from the density function given in Example 7.5, we note that integration by parts with $g(y) = y^{n-1}/(n-1)!$ and $h'(y) = \lambda^n e^{-\lambda x}$ gives

$$\int_t^\infty \frac{y^{n-1}\lambda^n}{(n-1)!} e^{-\lambda y} \, dy = \frac{y^{n-1}}{(n-1)!}(-\lambda^{n-1}e^{-\lambda y})\Big|_t^\infty$$
$$+ \int_t^\infty \frac{y^{n-2}\lambda^{n-1}}{(n-2)!} e^{-\lambda y} \, dy$$

So

$(\star)$ $$P(T_n > t) = \frac{\lambda^{n-1}t^{n-1}}{(n-1)!} e^{-\lambda t} + P(T_{n-1} > t)$$

The case $n = 1$ is the exponential distribution, so Example 2.7 implies $P(T_1 > t) = e^{-\lambda t}$. Using this in $(\star)$, we get

$$P(T_2 > t) = \lambda t e^{-\lambda t} + e^{-\lambda t}$$
$$P(T_3 > t) = \frac{(\lambda t)^2}{2!} e^{-\lambda t} + \lambda t e^{-\lambda t} + e^{-\lambda t}$$
$$P(T_n > t) = \sum_{m=0}^{n-1} e^{-\lambda t} \frac{(\lambda t)^m}{m!}$$

If we let Z be Poisson(λt) then the right-hand side is $P(Z < n)$ as claimed. □

A remarkable property of the Poisson process is

(7.5) If $s < t$ then N_s and $N_t - N_s$ are independent.

The proof is too difficult to give here. The key idea is that the lack of memory property of the exponential implies that the amount of time we have to wait for the first arrival after s, $T_{N_s+1} - s$, has an exponential distribution and is independent of the number of arrivals that have occurred by time s.

*3.8. Conditional Distributions

For discrete random variables, the definition of conditional probability implies

$$(8.1) \qquad P(X = x | Y = y) = \frac{P(X = x, Y = y)}{P(Y = y)} = \frac{P(X = x, Y = y)}{\sum_u P(X = u, Y = y)}$$

If we fix y and look at $P(X = x | Y = y)$ as a function of x, what we have is the **conditional distribution of X given that $Y = y$.**

Example 8.1. To illustrate this formula we look at our discrete example:

X	Y=100	125	150	175	
5	.08	.08	.06	0	.22
5.5	.08	.16	.16	.08	.48
6	0	.08	.10	.12	.30
	.16	.32	.32	.20	

In this case

$$P(X = 5 | Y = 150) = \frac{P(X = 5, \ Y = 150)}{P(Y = 150)} = \frac{.06}{.32} = 3/16$$

$$P(X = 5.5 | Y = 150) = \frac{P(X = 5.5, Y = 150)}{P(Y = 150)} = \frac{.16}{.32} = 1/2$$

$$P(X = 6 | Y = 150) = \frac{P(X = 6, \ Y = 150)}{P(Y = 150)} = \frac{.10}{.32} = 5/16$$

Or, in words, we take the third column and divide by the sum to make it a probability distribution.

Example 8.2. Suppose $X = \text{binomial}(n, p)$ and $Y = \text{binomial}(m, p)$ are independent. Example 7.1 implies that $X + Y = \text{binomial}(n + m, p)$. Since $P(X = j, X + Y = k) = P(X = j, Y = k - j) = P(X = j)P(Y = k - j)$

$$P(X = j | X + Y = k) = \frac{P(X = j)P(Y = k - j)}{P(X + Y = k)}$$

$$= \frac{C_{n,j} \, p^j (1 - p)^{n-j} \ C_{m,(k-j)} \, p^{k-j} (1 - p)^{m-(k-j)}}{C_{n+m,k} \, p^k (1 - p)^{n+m-k}}$$

$$= \frac{C_{n,j} \, C_{m,k-j}}{C_{n+m,k}}$$

This answer is already striking since it does not depend on p, but we can go further and rewrite it as something we can easily recognize. Using a trick that we have used several times in Chapter 2, we can write

$$\frac{C_{n,j}\, C_{m,k-j}}{C_{n+m,k}} = \frac{\frac{n!}{j!\,(n-j)!}\,\frac{m!}{(k-j)!\,(m-(k-j))!}}{\frac{(n+m)!}{k!\,(n+m-k)!}}$$

$$= \frac{\frac{k!}{j!\,(k-j)!}\,\frac{(n+m-k)!}{(n-j)!\,(m-(k-j))!}}{\frac{(n+m)!}{n!\,m!}} = \frac{C_{k,j}\, C_{n+m-k,n-j}}{C_{n+m,n}}$$

To see what this says, think of the k successes as green balls and the $n + m - k$ failures as red balls. Since all trials play the same role in $X + Y$, the conditional distribution of X given that $X + Y = k$ is the same as the distribution of the number of green balls we get when we draw n balls out of an urn with k green balls and $n + m - k$ red balls.

Formula (8.1) generalizes in the usual way to continuous distributions: on the right-hand side we replace $P(X = x, Y = y)$ by the joint density and $P(Y = y)$ by the marginal density of Y. Introducing $f_X(x|Y = y)$ as notation for the **conditional density** of X **given** $Y = y$ (which we think of as $P(X = x|Y = y)$), we have

$$(8.2) \qquad f_X(x|Y = y) = \frac{f(x,y)}{f_Y(y)} = \frac{f(x,y)}{\int f(u,y)\,du}$$

In words, we fix y, consider the joint density function as a function of x, and then divide by the integral to make it a probability density. To see how formula (8.2) works, we look at our continuous example.

Example 8.3.

$$f(x,y) = \begin{cases} e^{-y} & 0 < x < y < \infty \\ 0 & \text{otherwise} \end{cases}$$

In this case we have computed $f_Y(y) = ye^{-y}$ (in Example 5.2) so

$$f_X(x|Y = y) = \frac{e^{-y}}{ye^{-y}} = \frac{1}{y} \quad \text{for } 0 < x < y$$

That is, the conditional distribution is uniform on $(0, y)$. This should not be surprising since the joint density does not depend on x.

To compute the other conditional distribution we recall $f_X(x) = e^{-x}$ so

$$f_Y(y|X = x) = \frac{e^{-y}}{e^{-x}} = e^{-(y-x)} \quad \text{for } y > x$$

That is, given $X = x$, $Y - x$ is exponential with parameter 1. The last answer is quite reasonable since in Example 6.1 we saw that if Z_1, Z_2 are independent exponential(1) then $X = Z_1$, $Y = Z_1 + Z_2$ has the joint distribution given above. If we condition on $X = x$ then $Z_1 = x$ and $Y = x + Z_2$.

The multiplication rule says

$$P(X = x, Y = y) = P(X = x)P(Y = y|X = x)$$

Substituting in the analogous continuous quantities, we have

(8.3) $$f(x, y) = f_X(x)f_Y(y|X = x)$$

The next example demonstrates the use of (8.3) to compute a joint distribution.

Example 8.4. Suppose we pick a point uniformly distributed on $(0, 1)$, call it X, and then pick a point Y uniformly distributed on $(0, X)$. To find the joint density of (X, Y) we note that

$$\begin{aligned} f_X(x) &= 1 && \text{for } 0 < x < 1 \\ f_Y(y|X = x) &= 1/x && \text{for } 0 < y < x \end{aligned}$$

So using (8.3), we have

$$f(x, y) = f_X(x)f_Y(y|X = x) = 1/x \quad \text{for } 0 < y < x < 1$$

To complete the picture we compute

$$f_Y(y) = \int f(x, y)\, dx = \int_y^1 \frac{1}{x}\, dx = -\ln y$$

$$f_X(x|Y = y) = \frac{f(x, y)}{f_Y(y)} = \frac{1/x}{-\ln y} \quad \text{for } y < x < 1$$

Again the conditional density of X given $Y = y$ is obtained by fixing y, regarding the joint density function as a function of x, and then normalizing so that the integral is 1. The reader should note that although X is uniform on $(0, 1)$ and Y is uniform on $(0, X)$, X is not uniform on $(Y, 1)$ but has a greater probability of being near Y.

EXERCISE 8.1. Find the joint density that has the property that $f_X(x|Y = y)$ is uniform on $(y, 1)$ and $f_Y(y|X = x)$ is uniform on $(0, x)$.

The concept of conditional distribution generalizes easily to $n > 2$ variables. For example, in the discrete case, when $n = 4$,

$$P(X_2 = x_2, X_4 = x_4|X_1 = x_1, X_3 = x_3)$$

$$= \frac{P(X_1 = x_1, X_2 = x_2, X_3 = x_3, X_4 = x_4)}{P(X_1 = x_1, X_3 = x_3)}$$

$$= \frac{P(X_1 = x_1, X_2 = x_2, X_3 = x_3, X_4 = x_4)}{\sum_{y_1, y_3} P(X_1 = y_1, X_2 = x_2, X_3 = y_3, X_4 = x_4)}$$

while the analogous formula for the continuous case is

$$f_{X_2, X_4}(x_2, x_4 | X_1 = x_1, X_3 = x_3) = \frac{f(x_1, x_2, x_3, x_4)}{f_{X_1, X_3}(x_1, x_3)}$$
$$= \frac{f(x_1, x_2, x_3, x_4)}{\iint f(y_1, x_2, y_3, x_4) \, dy_1 \, dy_3}$$

EXERCISES

8.2. Compute (a) $P(X = 1 | Y = 1)$, (b) $P(X = 2 | Y = 2)$ for the following joint distribution:

Y	X=1	2	3
1	.1	.2	.3
2	.15	.15	0
3	.05	0	.05

8.3. Compute (a) $P(X = 2 | Y = 3)$, (b) $P(Y = 3 | X = 3)$ for the following joint distribution

Y	X=1	2	3
1	.2	.15	.05
2	.10	0	.10
3	.05	.15	.20

8.4. Using the clues given below, fill in the rest of the joint distribution. There is only one answer.

Y	X=0	3	6
1	?	?	?
2	.1	.05	?

(a) $P(Y = 2 | X = 0) = 1/4$, (b) X and Y are independent.

8.5. Using the clues given below, fill in the rest of the joint distribution. There is only one answer.

Y	X=1	2	3
1	?	?	?
2	?	0	?
3	0	?	0

For $k = 1, 2, 3$, (a) $P(Y = 1 | X = k) = 2/3$, (b) $P(X = k | Y = 1) = k/6$.

8.6. Fill in the rest of the joint distribution so that X and Y are independent. There are two possible answers

Y	X=0	1
0	?	2/9
1	2/9	?

8.7. Suppose we take a die with 3 on three sides, 2 on two sides, and 1 on one side, roll it n times, and let X_i be the number of times side i appeared. Find the conditional distribution $P(X_2 = k | X_3 = m)$.

8.8. Suppose $X = \text{Poisson}(\lambda)$ and $Y = \text{Poisson}(\mu)$ are independent. Find $P(X = m | X + Y = n)$.

8.9. Suppose X and Y are independent and have a geometric distribution with parameter p. Find $P(X = m | X + Y = n)$.

8.10. Suppose we observe a random variable X that has a Poisson distribution with parameter λ and then perform X independent trials with success probability p. Show that the number of successes we obtain has a Poisson distribution parameter λp.

8.11. Suppose $X_1, \ldots, X_m$ are independent and have a geometric distribution with parameter p. Find $P(X_1 = k | X_1 + \cdots + X_m = n)$.

8.12. Suppose $X = \text{gamma}(n, \lambda)$ and $Z = \text{gamma}(m, \lambda)$ are independent and let $Y = X + Z$. Find the conditional density of X given $Y = y$.

8.13. Suppose X and Y have joint density $f(x, y) = x + y$ when $0 < x < 1$ and $0 < y < 1$. Find the marginal density of X and the conditional density of Y given $X = x$.

8.14. Suppose X and Y have joint density $f(x, y) = 6x$ when $x, y > 0$ and $x + y < 1$. Find the marginal density of X and the conditional density of Y given $X = x$.

8.15. Suppose X and Y have joint density $f(x, y) = 8xy$ when $0 < y < x < 1$. Find the marginal density of X and the conditional density of Y given $X = x$.

8.16. Suppose X and Y have joint density $f(x, y) = (3x^2 + 4xy)/2$ when $0 < x, y < 1$. Find the marginal density of X and the conditional density of Y given $X = x$.

3.9. Chapter Summary and Review Problems

Section 3.1. The distribution of a discrete random variable is described by giving $P(X = x)$ for all values of x. Four famous discrete distributions are the hypergeometric, geometric, binomial, and Poisson.

Poisson Approximation to the Binomial. Suppose S_n has a binomial distribution with parameters n and p_n. If $p_n \to 0$ and $np_n \to \lambda$ as $n \to \infty$ then

$$(1.2) \qquad P(S_n = k) \to e^{-\lambda} \frac{\lambda^k}{k!}$$

Less formally, if n is large and p is small, binomial$(n, p) \approx$ Poisson$(\lambda = np)$.

Section 3.2. The distribution of a continuous random variable is described by giving its **density function** $f(x)$, which has

$$(2.2) \qquad\qquad f(x) \geq 0 \qquad \int f(x)\, dx = 1$$

We think of $f(x)$ as $P(X = x)$ because for all $a \leq b$

$$(2.1) \qquad\qquad P(a \leq X \leq b) = \int_a^b f(x)\, dx$$

Four famous continuous distributions are the uniform, exponential, gamma, and normal distributions.

Any random variable X (discrete or continuous) has a **distribution function** defined by $F(x) = P(X \leq x)$ and having the property that

$$(2.3) \qquad\qquad P(a < X \leq b) = F(b) - F(a)$$

In the continuous case $F(x) = \int_{-\infty}^x f(y)\, dy$. So F is the antiderivative of f with $F(-\infty) = 0$, and the last result is just the fundamental theorem of calculus:

$$\int_a^b f(x)\, dx = F(b) - F(a)$$

Since F is an antiderivative of f, we can recover the density function from the distribution function by differentiating:

$$f(x) = F'(x)$$

Intuitively, the median is the place where the distribution function F crosses $1/2$. Formally, m is a **median** if $P(X \leq m) \geq 1/2$ and $P(X \geq m) \geq 1/2$.

The distribution function is somewhat messy in the discrete case and it is usually better to work directly with the probability function $P(X = x)$.

Section 3.3. Suppose X has density function f and $P(a < X < b) = 1$. Let $Y = r(X)$. Suppose that $r : (a, b) \to (\alpha, \beta)$ is increasing and let $s : (\alpha, \beta) \to (a, b)$ be its inverse. Then Y has density

$$(3.1) \qquad\qquad g(y) = f(s(y))s'(y) \qquad \text{for } y \in (\alpha, \beta)$$

Two more special but also useful facts are:

(3.2) If X is a continuous random variable with distribution function F then $F(X)$ is uniform on (0,1).

(3.3) If U is uniform on (0,1) and $F^{-1}(y) = \min\{x : F(x) \leq y\}$ then $F^{-1}(U)$ has distribution function F.

Section 3.4. The joint distribution of X and Y is described in the discrete case by giving $P(X = x, Y = y)$ for all values of x and y, and in the continuous case by the **joint density function** $f(x,y)$, which has

(4.2) $$f(x,y) \geq 0 \qquad \iint f(x,y)\,dy\,dx = 1$$

We think of $f(x,y)$ as $P(X = x, Y = y)$ because

(4.1) $$P((X,Y) \in A) = \iint_A f(x,y)\,dy\,dx$$

The notion of a joint distribution extends in a straightforward way to $n > 2$ variables, but for this review we will stick to the case of two variables.

The **joint distribution function** is defined by $F(x,y) = P(X \leq x, Y \leq y)$. In the continuous case,

$$F(x,y) = \int_{-\infty}^{x} \int_{-\infty}^{y} f(u,v)\,dv\,du$$

or, to go in the other direction,

(4.5) $$\partial^2 F/\partial x \partial y = f(x,y)$$

We compute the probability (X,Y) lies in a rectangle by the formula

(4.4) $$P(a_1 < X \leq b_1, a_2 < Y \leq b_2)$$
$$= F(b_1, b_2) - F(a_1, b_2) - F(b_1, a_2) + F(a_1, a_2)$$

which is the two-dimensional analogue of $P(a < X \leq b) = F(b) - F(a)$.

Section 3.5. To recover the distribution of X from the joint distribution of X and Y, we use

(5.1) $$P(X = x) = \sum_{y} P(X = x, Y = y)$$

(5.2) $$f_X(x) = \int f(x,y)\,dy$$

Here and in similar instances below, the formula for the continuous case is obtained from the discrete case by writing density functions in place of the corresponding probabilities and replacing the sum by an integral.

Two random variables X and Y are said to be **independent** if

(5.3) $P(X \in A, Y \in B) = P(X \in A)P(Y \in B)$ for all sets A, B

In the discrete case this is equivalent to

(5.4) $P(X = x, Y = y) = P(X = x)P(Y = y)$ for all x, y

In the continuous case this is equivalent to

(5.5) $f(x,y) = f_X(x)f_Y(y)$ for all x, y

or to being able to write

(5.6) $f(x,y) = g(x)h(y)$ for all x, y

Recall that X and Y are not independent unless $\{(x,y) : f(x,y) > 0\}$ is a rectangle.

If $X_1, \ldots, X_{n+m}$ are independent then

(5.7) $r(X_1), \ldots, r_{n+m}(X_{n+m})$ are independent

(5.8) $f(X_1, \ldots, X_n)$ and $g(X_{n+1}, \ldots, X_{n+m})$ are independent

Section 3.6. Let S be a subset of R^n so that $P((X_1, \ldots, X_n) \in S) = 1$ and suppose that the function r maps S 1-1 onto a set T. That is, $\{r(x) : x \in S\} = T$ and there is an inverse function s from T to S, so that $s(r(x)) = x$ for $x \in S$. Let s_i be the ith component of the inverse function, let $D_{ij} = \partial s_i / \partial y_j$ be the matrix of partial derivatives, and let $J = \det D$ be its determinant. J is for Jacobian. The density function of $Y = r(X)$ is given by

(6.1) $$f_Y(y) = \begin{cases} f_X(s(y))|J| & \text{for } y \in T \\ 0 & \text{otherwise} \end{cases}$$

Section 3.7. If X and Y are independent then

(7.1) $$P(X + Y = z) = \sum_x P(X = x)P(Y = z - x)$$

(7.2) $$f_{X+Y}(z) = \int f_X(x)f_Y(z - x)\,dx$$

Some important special cases (and their example numbers) are

		X	Y	$X+Y$
7.1.	binomial	(m,p)	(n,p)	$(m+n,p)$
7.5.	gamma	(m,λ)	(n,λ)	$(m+n,\lambda)$
7.2.	Poisson	μ	ν	$\mu+\nu$
7.4.	normal	(μ,a)	(ν,b)	$(\mu+\nu,a+b)$

The first relationship holds because the number of successes in n independent trials with success probability p is binomial(n,p). The second holds because the sum of k independent exponential(λ) random variables is gamma(k,λ).

Section 3.8. The conditional distribution of X given that $Y=y$ is

$$(8.1) \quad P(X=x|Y=y) = \frac{P(X=x,Y=y)}{P(Y=y)} = \frac{P(X=x,Y=y)}{\sum_u P(X=u,Y=y)}$$

$$(8.2) \qquad f_X(x|Y=y) = \frac{f(x,y)}{f_Y(y)} = \frac{f(x,y)}{\int f(u,y)\,du}$$

Formula (8.1) is a consequence of the definition of conditional probability: $P(B|A) = P(A \cap B)/P(A)$. The next formula comes from the multiplication rule: $P(A \cap B) = P(A)P(B|A)$.

$$(8.3) \qquad \begin{aligned} P(X=x,Y=y) &= P(X=x)P(Y=y|X=x) \\ f(x,y) &= f_X(x)f_Y(y|X=x) \end{aligned}$$

EXERCISES

9.1. Suppose we roll two dice that have 3 on three sides, 2 on two sides, and 1 on one side. Find the probability function for the sum of the two numbers that appear.

9.2. Suppose we draw 3 balls out of an urn with 5 red and 4 black balls. Find the probability function for the number of red balls drawn.

9.3. How many children should a family plan to have so that the probability of having at least one child of each sex is at least 0.95?

9.4. Use the Poisson approximation to compute the probability that you will roll at least one double 6 in 24 trials. How does this compare with the exact answer?

9.5. Suppose that each student in a freshman class of 1000 has probability 1/2000 of commiting suicide in their freshman year. Use the Poisson approximation to compute the probability there will be four or more suicides by freshmen.

9.6. Suppose X has density function $2 - 2x$ for $0 < x < 1$ and 0 otherwise. Find (a) the distribution function, (b) $P(X > 1/2)$, and (c) the median of X.

9.7. Suppose X has density function $x^{-1/2}/2$ for $0 < x < 1$ and 0 otherwise. Find (a) the distribution function, (b) $P(X < 1/9)$, and (c) the median of X.

9.8. Suppose X has density function $f(x) = e^x/(1 + e^x)^2$. Find (a) the distribution function, (b) $P(X > \ln 2)$.

9.9. Let $F(x) = 1 - x^4 e^{-x^2}$ for $x \geq 0$. Is this a distribution function? If so, find its density function.

9.10. Find the constant c to make $f(x) = ce^{-x^2-2x}$ a density function.

9.11. Suppose $X = $ binomial$(4, 0.4)$. Find all of the medians of X.

9.12. A number U is picked according to a uniform distribution on $(0,1)$. What is the probability that the first digit in the decimal expansion of $\sqrt{U}$ is k?

9.13. An absent-minded person puts an exponential(1) amount of air in a balloon. Find the distribution of the radius of the balloon. Recall that the volume of a sphere of radius r is $V = 4\pi r^3/3$.

9.14. Pick a point at random from the sphere of radius 1. Find the distribution function for $R = $ the distance from the center.

9.15. Suppose X has a normal distribution and $Y = X^2$. (a) Find the density function of Y, which is called the **chi-square distribution**. (b) If we let $Y_1, \ldots, Y_m$ be independent and have the same distribution as Y, the sum is called a **chi-square distribution with m degrees of freedom**. What is its density function?

9.16. Suppose $X = $ normal(0,1) and $Y = $ gamma$(m/2, 1/2)$ are independent. Find the density function of $Z = x/\sqrt{Y/m}$ This is a t **distribution with m degrees of freedom**.

9.17. Suppose X and Y are independent and have exponential(λ) distributions. Find the density function of $X - Y$.

9.18. Suppose X and Y are independent exponential(1). Find $P(X \geq Y \geq 2)$.

9.19. Two people have agreed to meet at a party. If we call the start of the party time 0, then Mary (who wants to be late but not too late) arrives at time X (measured in hours) with density

$$f_X(x) = 1 \qquad \text{for } 0 < x < 1$$

while John (the space cadet) arrives at time Y (again measured in hours) with density

$$f_Y(y) = e^{-y} \qquad \text{for } y > 0$$

Suppose (a) the arrival times X and Y are independent and (b) Mary will get impatient and leave if John arrives more than one hour after she does. What is the probability Mary will leave before John arrives, i.e., what is $P(Y > X+1)$?

9.20. Suppose (X, Y) is uniformly distributed over the set where $0 \le y \le 1-x^2$ and $-1 \le x \le 1$. Find the marginal densities of X and Y.

9.21. Let $F(x, y) = 1 - e^{-xy}$ when $x > 0$ and $y > 0$. Is F a joint distribution function? If so, find its density function.

9.22. Suppose X and Y have joint density function

$$f(x,y) = \begin{cases} 60x^2y & \text{when } x > 0,\ y > 0,\ x+y < 1 \\ 0 & \text{otherwise} \end{cases}$$

(a) Find the marginal density of X. (b) Find the conditional density of Y given $X = x$.

9.23. Suppose X and Y have joint density function

$$f(x,y) = \begin{cases} 6(x - y) & \text{when } 0 < y < x < 1 \\ 0 & \text{otherwise} \end{cases}$$

(a) Find the marginal density of X. (b) Find the conditional density of Y given $X = x$.

9.24. Suppose X and Y have joint density $f(x, y) = (1/2)e^{-y}$ when $y \ge 0$ and $-y \le x \le y$. Compute $P(X \le 1 | Y = 3)$.

9.25. A circuit board has three wires and will continue to function as long as two of the wires are working. Suppose that the times to failure of the three wires are independent and exponential(1). Find the density function of the time T to circuit board failure.

9.26. Jobs 1 and 2 must be completed before job 3 is begun. If the amount of time each task takes is independent and uniform on $(2,4)$, find the density function for the amount of time T it takes to complete all three jobs.

4 Expected Value

4.1. Examples

In this section we will introduce the expected value or mean of a random variable X. This quantity has a meaning much like the frequency interpretation of probability. In Chapter 5 we will see that if $X_1, \ldots, X_n$ are independent and have the same distribution as X then, when n is large, the average of the values we have observed, $(X_1 + \cdots + X_n)/n$, will be close to EX with high probability. This result is called the **law of large numbers**.

If X has a discrete distribution then the **expected value** of X is

$$(1.1) \qquad EX = \sum_x x P(X = x)$$

Example 1.1. Roll one die and let X be the number that appears. $P(X = x) = 1/6$ for $x = 1, 2, 3, 4, 5, 6$ so

$$EX = 1 \cdot \frac{1}{6} + 2 \cdot \frac{1}{6} + 3 \cdot \frac{1}{6} + 4 \cdot \frac{1}{6} + 5 \cdot \frac{1}{6} + 6 \cdot \frac{1}{6} = \frac{21}{6} = 3\frac{1}{2}$$

In this case the expected value is just the average of the six possible values.

Example 1.2. Roulette. If you play roulette and bet $1 on black then you win $1 with probability 18/38 and you lose $1 with probability 20/38, so the expected value of your winnings X is

$$EX = 1 \cdot \frac{18}{38} + (-1) \cdot \frac{20}{38} = \frac{-2}{38} = -0.0526$$

If you play n times and let X_i be your winnings on the ith play then the law of large numbers implies that $(X_1 + \cdots + X_n)/n$ will be close to -0.0526. In words, in the long run you will lose about 5.26 cents per play.

Example 1.3. A two-person game. After counting "one, two, three," each player shows one or two fingers. If the sum is odd (i.e., 3) the first player pays the second \$3. If the sum is even (i.e., 2 or 4) the second player pays the first the sum. If the second player shows one finger with probability p and two fingers with probability $1 - p$ then the first player wins an average of $2p + (-3)(1 - p)$ by showing one finger and an average of $(-3)p + 4(1 - p)$ by showing two fingers. Setting these two quantities equal, $5p - 3 = -7p + 4$, and solving gives $p = 7/12$. That is, if the second player shows one finger $7/12$ of the time and two fingers $5/12$ of the time then the first player's average payoff from showing one or two fingers is the same, $-1/12$. So no matter what player 1 does, player 2 will win an average of $1/12$ of a dollar per play. By symmetry, if player 1 shows one finger $7/12$ of the time and two fingers $5/12$ of the time, he will lose an average of $1/12$ of a dollar per play no matter what player 2 does.

Example 1.4. Keno. In Keno a player receives a card with the numbers 1 through 80 on it. She then marks the numbers she wants to play (anywhere from 1 to 15 of them) and indicates the amount of her bet. Twenty numbers are then drawn without replacement from the numbers 1 through 80, and the player is paid an amount that depends on how many marked numbers were chosen. For instance, if she bets \$.60 and picks 10 numbers she gets \$1.20 if 5 of her 10 numbers are chosen, \$12.00 if 6 are chosen, \$90 for 7, \$660 for 8, \$2,400 for 9, and \$12,500 if all 10 are chosen. To compute the expected value of this bet we begin by observing that the probability i of her numbers are chosen is $C_{10,i}C_{70,20-i}/C_{80,20}$ so her expected winnings are

$$EW = \sum_{i=5}^{10} \mathrm{pay}(i) \frac{C_{10,i}\,C_{70,20-i}}{C_{80,20}}$$

where $\mathrm{pay}(i)$ is the payoff when i of her numbers are chosen. Using a computer to evaluate the probabilities we arrive at the following

i	pay(i)	prob	pay(i) × prob
5	1.20	.051428	.06171
6	12.00	.011479	.13775
7	90.00	.001611	.14500
8	660.00	.000135	.08938
9	2,400.00	6.12×10^{-6}	.01469
10	12,500.00	1.12×10^{-7}	.00140
sum		.064660	.44994

As the bottom line indicates, the expected value of our bet is about 45 cents or 75% of our bet, and we win something about 6.46% of the time. All of the 15 Keno bets have about the same expected value but some pay off more frequently. Notice that if the payoff for getting all 10 was raised to $100,000 the expected value would only increase by

$$\$87,500 \times 1.12 \times 10^{-7} \approx 1 \text{ cent}$$

Example 1.5. Poisson distribution. Suppose $P(X = k) = e^{-\lambda}\lambda^k/k!$ for $k = 0, 1, 2, \ldots$ In this case since the $k = 0$ term makes no contribution to the sum,

$$EX = \sum_{k=1}^{\infty} k e^{-\lambda} \frac{\lambda^k}{k!} = \lambda \sum_{k=1}^{\infty} e^{-\lambda} \frac{\lambda^{k-1}}{(k-1)!} = \lambda$$

since $\sum_{k=1}^{\infty} P(X = (k-1)) = 1$.

Note. As we go along we will accumulate a number of facts about our most important distributions: binomial, hypergeometric, geometric, Poisson, uniform, exponential, gamma, and normal. For easy reference, these facts are collected together in a section at the end of the book.

The definition of expected value generalizes in the usual way to continuous random variables: We replace the probability function by the density function and the sum by an integral

$$(1.2) \qquad\qquad EX = \int x f(x)\, dx$$

Example 1.6. Uniform distribution on (a,b). Suppose X has density function $f(x) = 1/(b-a)$ for $a < x < b$ and 0 otherwise. In this case

$$EX = \int_a^b \frac{x}{b-a}\, dx = \frac{b^2}{2(b-a)} - \frac{a^2}{2(b-a)} = \frac{(b+a)}{2}$$

since $b^2 - a^2 = (b+a)(b-a)$. Notice that $(a+b)/2$ is the midpoint of the interval and hence the natural choice for the average value of X.

Example 1.7. Gamma distribution. Suppose X has density function $f(x) = \lambda^n x^{n-1} e^{-\lambda x}/(n-1)!$ for $x \geq 0$ and 0 otherwise. To compute EX we manipulate the integrand to make it look like the gamma$(n+1, \lambda)$ density, which integrates to 1.

$$EX = \int_0^{\infty} x \frac{\lambda^n x^{n-1}}{(n-1)!} e^{-\lambda x}\, dx$$

$$= \frac{n}{\lambda} \int_0^{\infty} \frac{\lambda^{n+1} x^n}{n!} e^{-\lambda x}\, dx = \frac{n}{\lambda}$$

When $n = 1$ this says that the exponential distribution with parameter λ has expected value $1/\lambda$.

In formulating the definition of expected value we have made a minor error of omission, which we will now correct. The expected value is only defined in the discrete case when $\sum_x |x|P(X = x) < \infty$ and in the continuous case when $\int |x|f(x)\,dx < \infty$. If the sum or integral is infinite we say that the expected value does not exist. This assumption is more than just a technicality. It is needed to guarantee that $(X_1 + \cdots + X_n)/n$ converges to EX as $n \to \infty$.

Our next three examples give situations in which $E|X| = \infty$.

Example 1.8. The Cauchy density. $f(x) = 1/\pi(1 + x^2)$. In this case the expected value does not exist since

$$E|X| = \int \frac{|x|}{\pi(1 + x^2)}\,dx = \infty$$

The intuition behind the last conclusion is that when x is large the 1 in the denominator is insignificant, so the integrand is approximately $1/\pi x$, which is not integrable. To translate the intuition into a proof we observe that

$$E|X| \geq \int_1^\infty x \frac{1}{\pi(1 + x^2)}\,dx$$
$$\geq \int_1^\infty \frac{x}{\pi(2x^2)}\,dx = \frac{1}{2\pi}\int_1^\infty \frac{1}{x}\,dx = \infty$$

the second inequality holding since $1 + x^2 \leq 2x^2$ for $x \geq 1$.

One might be tempted to argue that "by symmetry" the Cauchy density must have $EX = 0$, but this conclusion does not agree with our interpretation of EX. When $X_1, \ldots, X_n$ are independent and have a Cauchy distribution, the average of the first n values $(X_1 + \cdots + X_n)/n$ does not settle down to a limit. Indeed, the Cauchy distribution has the very special property that for all n the average $(X_1 + \cdots + X_n)/n$ has the same distribution as X_1.

It is natural to think that examples in which the mean does not exist are pathological. Our next two examples shows that such distributions can arise in "real life."

Example 1.9. Suppose we flip a coin and keep track of $H_n =$ the number of Heads in the first n tosses and $T_n = n - H_n =$ the number of Tails. Let

$$N = \min\{n : H_n = T_n\}$$

be the first time the numbers of Heads and Tails are equal. This can only happen when n is even. A calculation that we will give at the end of Section 4.2 shows

$$(\star) \qquad P(N = 2m) = \frac{1}{2m-1}\binom{2m}{m}2^{-2m}$$

To see what $P(N = 2m)$ looks like for large m we will use Stirling's formula:

$$n! \sim n^n e^{-n}\sqrt{2\pi n}$$

where $a_n \sim b_n$ means that $a_n/b_n \to 1$ as $n \to \infty$. Using this and noticing $e^{-2m}/e^{-m}e^{-m} = 1$, we have

$$2^{-2m}\frac{(2m)!}{m!\,m!} \sim 2^{-2m}\frac{(2m)^{2m}}{(m^m)^2}\frac{\sqrt{4\pi m}}{(\sqrt{2\pi m})^2} = 1/\sqrt{\pi m}$$

Since $(2m-1) \sim 2m$ it follows that

$$P(N = 2m) \sim (1/2\sqrt{\pi})m^{-3/2}$$

The last equation implies $2mP(N = 2m) \sim (1/\sqrt{\pi})m^{-1/2}$ so

$$\sum_{m=0}^{\infty} 2mP(N = 2m) = \infty$$

To compute the expected value in our next example, we need a fact that is useful in several situations.

(1.3) If $X \geq 0$ is integer-valued then

$$EX = \sum_{n=1}^{\infty} P(X \geq n)$$

PROOF: Since $\{X \geq n\} = \cup_{m=n}^{\infty}\{X = m\}$ and the events on the right-hand side are disjoint,

$$P(X \geq n) = \sum_{m=n}^{\infty} P(X = m)$$

Summing the last equality from $n = 1$ to ∞ we have

$$\sum_{n=1}^{\infty} P(X \geq n) = \sum_{n=1}^{\infty}\sum_{m=n}^{\infty} P(X = m)$$

Interchanging the order of summation, the last expression

$$= \sum_{m=1}^{\infty} \sum_{n=1}^{m} P(X = m) = \sum_{m=1}^{\infty} m P(X = m) = EX$$

since there is no contribution to the expectation from the $m = 0$ term. $\square$

Example 1.10. You are selling a car. Let $X_0, X_1, \ldots$ be the successive offers you receive. Suppose that they are independent and have the same continuous distribution. Let $N = \min\{n : X_n > X_0\}$ be the number of additional offers you need to get one better than the first. Now

$$P(N \geq n) = P(X_0 \geq X_1, X_0 \geq X_2, \ldots, X_0 \geq X_{n-1}) = 1/n$$

since there are n random variables and by symmetry each has the same probability of being the largest. (We have assumed that they have a continuous distribution so that the probability that two are equal is 0.) Summing and using (1.3), we have

$$EN = \sum_{n=1}^{\infty} P(N \geq n) = \sum_{n=1}^{\infty} \frac{1}{n} = \infty$$

so you expect to wait a long time to see an offer better than the first one.

EXERCISE 1.1. Show that for any distribution $P(N \geq n) \geq 1/n$ and hence $EN = \infty$.

The formula in (1.3) is also useful for

Example 1.11. Geometric distribution. Let N be the number of independent trials we need to obtain a success when success has probability p. In this case $P(N \geq n) = (1 - p)^{n-1}$, since $N \geq n$ if and only if the first $n - 1$ trials result in a failure. Using (1.3) now,

$$EN = \sum_{n=1}^{\infty} (1 - p)^{n-1} = 1/p$$

since (set $x = 1 - p$, $m = n - 1$)

(1.4)
$$(1 - x)^{-1} = \sum_{m=0}^{\infty} x^m$$

This answer is quite reasonable: the average number of successes on one trial is p, so the average number of trials needed for one success is $1/p$.

EXERCISES

1.2. You want to invent a gambling game in which a person rolls two dice and is paid some money if the sum is 7, but otherwise he loses his money. How much should you pay him for winning a $1 bet if you want this to be a fair game, that is, to have expected value 0?

1.3. A bet is said to carry 3 to 1 odds if you win $3 for each $1 you bet. What must the probability of winning be for this to be a fair bet?

1.4. A roulette wheel has slots numbered 1 to 36 and two labeled with 0 and 00. Suppose that all 38 outcomes have equal probability. Compute the expected values of the following bets. In each case you bet one dollar and when you win you get your dollar back in addition to your winnings. (a) You win $1 if one of the numbers 1 through 18 comes up. (b) You win $2 if the number that comes up is divisible by 3 (0 and 00 do not count). (c) You win $35 if the number 7 comes up.

1.5. In the Las Vegas game Wheel of Fortune, there are 54 possible outcomes. One is labeled "Joker," one "Flag," two "20," four "10," seven "5," fifteen "2," and twenty-four "1." If you bet $1 on a number you win that amount of money if the number comes up (plus your dollar back). If you bet $1 on Flag or Joker you win $40 if that symbol comes up (plus your dollar back). What bets have the best and worst expected value here?

1.6. A person playing Keno picks 3 numbers and bets $0.60. She wins $ 26 if all three numbers are chosen, $ 0.60 if two are, and $0 otherwise. Compute the expected value of her winnings.

1.7. In blackjack the dealer gets two cards, one of which you can see and one of which you cannot. When the dealer's visible card is an Ace, she offers you a chance to take out "insurance." You can bet $1 that the invisible card is a face card or a 10. If it is, you win $2, otherwise you lose $1. What is the expected value of this bet (a) if we assume that the dealer's other card was chosen at random from a second deck of 52? (b) if we use the information that the dealer's Ace and our two cards, which are a 6 and an 8, came from the same deck of 52?

1.8. Suppose we flip a coin 4 times and let X be the number of heads that appear. Find EX by computing $\sum_x x P(X = x)$.

1.9. Suppose we roll two dice and let X be the sum of the two dice. Find EX by computing $\sum_x x P(X = x)$.

1.10. Every Sunday afternoon two children, Al and Betty, take turns trying to fly their kites. Suppose Al is successful with probability 1/5 and Betty is successful with probability 1/3. How many Sundays do we expect to have to wait until at least one kite flies?

1.11. Suppose we roll a die repeatedly and let T_k be the number of rolls until some number appears k times in a row. (a) Compute ET_2. (b) Find a formula that relates ET_k to ET_{k-1}. (c) Compute ET_3.

1.12. Suppose X has density function $f(x) = 4x^3$ for $0 < x < 1$, 0 otherwise. Find EX.

1.13. Suppose X has density function $f(x) = 3x(2 - x)/4$ for $0 < x < 2$, 0 otherwise. Find EX.

1.14. Suppose X has the standard normal distribution. Compute $E|X|$.

1.15. Suppose X has a density function f with $f(c + x) = f(c - x)$ for all $x \geq 0$. That is, f is symmetric about c. Show that if $\int |x| f(x)\, dx < \infty$ then $\int x f(x)\, dx = c$.

1.16. Suppose X has the power law density, i.e., $f(x) = (\rho - 1)x^{-\rho}$ for $x \geq 1$ and 0 otherwise. Show that $EX = \infty$ when $\rho \leq 2$ and find EX for $\rho > 2$.

1.17. Suppose that X has density function $cx^{-2}(\ln x)^{-a}$ for $x \geq 1$, 0 otherwise, where c is chosen to make this a probability density. Show that EX exists if $a > 1$ but not if $a \leq 1$.

1.18. Let $X_1, X_2, \ldots$ be independent and have a continuous distribution F. Let $N \geq 2$ be the first time $X_1 > X_2 \ldots > X_{N-1} < X_N$. That is the first time $X_1, \ldots, X_N$ fails to be a decreasing sequence. Find EN.

1.19. Suppose $X \geq 0$ has density function f. Use integration by parts to show that $EX = \int_0^\infty P(X > x)\, dx$. To do this it is useful to note that

$$x P(X > x) \leq \int_x^\infty y f(y)\, dy \to 0 \quad \text{as } x \to \infty$$

1.20. Suppose $X_1, \ldots, X_n \geq 0$ are independent and have distribution F. Use the formula in the last exercise to show

$$E \min\{X_1, \ldots, X_n\} = \int_0^\infty (1 - F(x))^n\, dx$$

$$E \max\{X_1, \ldots, X_n\} = \int_0^\infty 1 - F(x)^n\, dx$$

1.21. Suppose $U_1, \ldots, U_n$ are independent and uniform on $(0,1)$ and let $Y = \min\{U_1, \ldots, U_n\}$, $Z = \max\{U_1, \ldots, U_n\}$. Find (a) EY, (b) EZ.

1.22. A bowl contains balls numbered $1, \ldots, N$. Let X be the largest number obtained when n balls are drawn without replacement. Find an exact formula for EX and show that $EX \approx Nn/(n+1)$ when n is large. According to Feller, this method was used during World War II to estimate the enemy's production of weapons. For example if the largest serial number among 10 captured tanks was 1000 we would guess that

$$N = (n+1)X/n = 11(1000)/10 = 1100$$

1.23. Suppose $X_1, \ldots, X_n$ are independent and have an exponential distribution with parameter 1 and let $Y = \min\{X_1, \ldots, X_n\}$, $Z = \max\{X_1, \ldots, X_n\}$. Find (a) EY, (b) EZ. Hint for (b): relate the answer for n to the answer for $n - 1$.

1.24. Suppose we roll five dice. We leave the ones that come up 6 alone and roll the others again, repeating this procedure until all five dice show 6. Let R be the number of rolls required. Find ER.

4.2. Moments, Generating Functions

In this section we will look at the expected value of functions of a random variable. The first basic fact is

(2.1) If X has a discrete distribution and $Y = r(X)$ then

$$EY = \sum_x r(x)P(X = x)$$

PROOF: $P(Y = y) = \sum_{x:r(x)=y} P(X = x)$. Multiplying both sides by y and summing gives

$$EY = \sum_y y\, P(Y = y) = \sum_y y \sum_{x:r(x)=y} P(X = x)$$

$$= \sum_y \sum_{x:r(x)=y} r(x)P(X = x) = \sum_x r(x)P(X = x)$$

since the double sum is just a clumsy way of summing over all the possible values of x. □

Example 2.1. Suppose X is uniform on $\{-3, -2, -1, 0, 1, 2, 3\}$. Compute EX^2.

By the formula in (2.1),

$$EX^2 = \frac{1}{7}(9 + 4 + 1 + 0 + 1 + 4 + 9) = 28/7 = 4$$

To check the answer let $Y = X^2$ and note that $P(Y = 0) = 1/7$, $P(Y = i) = 2/7$ for $i = 1, 4, 9$ so

$$EY = 0 \cdot \frac{1}{7} + 1 \cdot \frac{2}{7} + 4 \cdot \frac{2}{7} + 9 \cdot \frac{2}{7} = \frac{28}{7} = 4$$

The analogue of (2.1) holds in the continuous case:

(2.2) If X has density f and $Y = r(X)$ then

$$EY = \int r(x) f(x) \, dx$$

PROOF: The proof for a general function r is too difficult to give here but we can easily prove this when r is increasing. Let s be the inverse of r, defined by $r(s(y)) = y$. Changing variables $x = s(y)$, $dx = s'(y) \, dy$ and then using (3.1) from Chapter 3 to recognize $f_X(s(y))s'(y)$ as the density of Y we have

$$\int r(x) f_X(x) \, dx = \int y f_X(s(y)) s'(y) \, dy = \int y f_Y(y) \, dy = EY \qquad \square$$

Example 2.2. Suppose X has density $f(x) = rx^{r-1}$ when $0 < x < 1$, and 0 otherwise. Here, $r > 0$. When $r = 1$, X is uniform on (0,1).

$$EX^k = \int_0^1 x^k \, rx^{r-1} \, dx = \frac{rx^{k+r}}{k+r} \Big|_0^1 = \frac{r}{k+r}$$

In general, EX^k is called the **kth moment** of X.

Example 2.3. Suppose X has an exponential density with parameter λ. Then

$$Ee^{tX} = \int_0^\infty e^{tx} \lambda e^{-\lambda x} \, dx = \lambda \int_0^\infty e^{-(\lambda - t)x} \, dx = \frac{\lambda}{\lambda - t}$$

for $t < \lambda$. $Ee^{tX} = \infty$ for $t \geq \lambda$. The last calculation extends easily to the gamma density. To compute Ee^{tX} for $t < \lambda$ we manipulate the integrand to make it look like a gamma density with parameters n and $\lambda-t$, which integrates to 1.

$$Ee^{tX} = \int e^{tx} \frac{\lambda^n x^{n-1}}{(n-1)!} e^{-\lambda x} \, dx$$

$$= \frac{\lambda^n}{(\lambda-t)^n} \int \frac{(\lambda-t)^n x^{n-1}}{(n-1)!} e^{-(\lambda-t)x} \, dx = \frac{\lambda^n}{(\lambda-t)^n}$$

The last formula is valid for $t < \lambda$. Again $Ee^{tX} = \infty$ for $t \geq \lambda$.

In general $\phi_X(t) = Ee^{tX}$ is called the **moment generating function** of X (or m.g.f. for short) because as we will now explain, it can be used to generate the moments of X.

(2.3) Suppose $\phi_X(t) = Ee^{tX} < \infty$ for $t \in (-t_0, t_0)$ for some $t_0 > 0$ and let $\phi^{(n)}$ denote the nth derivative of ϕ. Then

$$\phi^{(n)}(0) = EX^n$$

PROOF: Ignoring the detail of the interchange of the operations of differentiation and expectation (which is valid under the assumption $\phi_X(t) = Ee^{tX} < \infty$ for $t \in (-t_0, t_0)$ for some $t_0 > 0$) we have

$$\phi'_X(t) = \frac{d}{dt} E(e^{tX}) = E\left(\frac{d}{dt} e^{tX}\right) = E(Xe^{tX})$$

Setting $t = 0$ we have $\phi'_X(0) = EX$. Differentiating again,

$$\phi''_X(t) = \frac{d}{dt} \phi'_X(t) = \frac{d}{dt} E(Xe^{tX}) = E\left(\frac{d}{dt} Xe^{tX}\right) = E(X^2 e^{tX})$$

and setting $t = 0$ we have $\phi''_X(0) = EX^2$. Each differentiation brings down another factor of X and setting $t = 0$ makes $e^{tX} = 1$ so the general result follows.   □

As an application of the last formula we can compute the moments of the exponential distribution. To begin, we note that from Example 2.3, $\phi_X(t) = \lambda/(\lambda-t)$ so

$$\phi'_X(t) = \frac{\lambda}{(\lambda-t)^2} \qquad EX = \phi'_X(0) = 1/\lambda$$

$$\phi''_X(t) = \frac{2\lambda}{(\lambda-t)^3} \qquad EX^2 = \phi''_X(0) = 2/\lambda^2$$

$$\phi'''_X(t) = \frac{3 \cdot 2\lambda}{(\lambda-t)^4} \qquad EX^3 = \phi'''_X(0) = 3!/\lambda^3$$

From the first three terms we can see that in general

$$\phi_X^{(n)}(t) = \frac{n!\lambda}{(\lambda - t)^{n+1}} \qquad EX^n = \phi_X^{(n)}(0) = n!/\lambda^n$$

EXERCISE 2.1. Derive the last result by using integration by parts to establish that $EX^n = (n/\lambda)EX^{n-1}$.

EXERCISE 2.2. Differentiate the moment generating function of the gamma distribution to find its moments.

The moment generating function is defined for any random variable but if X is a nonnegative integer-valued random variable, it is often more convenient to look at its **generating function**, defined by

$$\gamma_X(z) = Ez^X = \sum_{k=0}^{\infty} z^k P(X = k)$$

Here we use the convention that $z^0 = 1$ even when $z = 0$. Note that if we set $z = e^t$ in the generating function we get the moment generating function

$$\gamma_X(e^t) = Ee^{tX} = \phi_X(t)$$

Example 2.4. Poisson distribution. If $P(X = k) = e^{-\lambda}\lambda^k/k!$ for $k = 0, 1, 2, \ldots$ then the generating function is given by

$$\gamma_X(z) = \sum_{k=0}^{\infty} e^{-\lambda}\frac{\lambda^k}{k!}z^k$$

$$= e^{-\lambda}\sum_{k=0}^{\infty}\frac{(\lambda z)^k}{k!} = e^{-\lambda}e^{\lambda z} = e^{\lambda(z-1)}$$

Setting $z = e^t$ we get the m.g.f. $\phi_X(t) = e^{\lambda(e^t-1)}$.

One of the reasons for interest in the generating function is that it can also be used to compute moments.

(2.4) Let X be a nonnegative integer-valued random variable, let $\gamma_X(z) = Ez^X$ be its generating function, and let $\gamma_X^{(n)}$ be the nth derivative of γ_X. Then

$$\gamma_X^{(n)}(1) = E\{X(X - 1)\cdots(X - n + 1)\}$$

The quantity on the right-hand side is sometimes called the **nth factorial moment**.

PROOF: Differentiating the definition (not worrying about interchanging the derivative and the sum, which is legitimate in this case) and then setting $z = 1$ we have

$$\gamma'(z) = \sum_{k=1}^{\infty} k z^{k-1} P(X = k)$$

$$\gamma'(1) = \sum_{k=1}^{\infty} k P(X = k) = EX$$

$$\gamma''(z) = \sum_{k=2}^{\infty} k(k-1) z^{k-2} P(X = k)$$

$$\gamma''(1) = \sum_{k=2}^{\infty} k(k-1) P(X = k) = E\{X(X-1)\}$$

Here we have removed values of k from the sum that correspond to terms that are 0. It should be clear from the first two formulas that in general

$$\gamma^{(n)}(z) = \sum_{k=n}^{\infty} k(k-1)\cdots(k-n+1) z^{k-n} P(X = k)$$

$$\gamma^{(n)}(1) = \sum_{k=n}^{\infty} k(k-1)\cdots(k-n+1) P(X = k)$$

$$= E\{X(X-1)\cdots(X-n+1)\} \qquad \square$$

Example 2.5. In the case of the Poisson: $\gamma(z) = e^{\lambda(z-1)}$, so

$$\gamma'(z) = \lambda e^{\lambda(z-1)} \qquad \gamma'(1) = \lambda$$

which agrees with the answer in Example 1.5. Continuing to differentiate, we get new information:

$$\gamma''(z) = \lambda^2 e^{\lambda(z-1)} \qquad \gamma''(1) = \lambda^2$$

It should be clear from the first two formulas that in general

$$\gamma^{(n)}(z) = \lambda^n e^{\lambda(z-1)} \qquad \gamma^{(n)}(1) = \lambda^n$$

so $E\{X(X-1)\cdots(X-n+1)\} = \lambda^n$.

EXERCISE 2.3. Differentiate the moment generating function of the Poisson to find EX and EX^2.

Our final topic is an optimization problem:

Example 2.6. A vendor in the lobby of a big office building sells coffee. Suppose that the demand for coffee, measured in gallons, has density function $f(x)$. Suppose that she must pay c dollars per gallon for coffee, receives $b > c$ dollars for each gallon sold, but suffers a penalty of a dollars per gallon in lost goodwill for unfulfilled demand. How much coffee should she buy to maximize her profit?

If she buys v gallons and the demand is X then her profit is

$$\begin{cases} (b-c)v - a(X-v) & \text{if } X \geq v \\ -cv + Xb = (b-c)v - b(v-X) & \text{if } X < v \end{cases}$$

So her expected profit is

$$(b-c)v - \int_v^\infty a(x-v)f(x)\,dx - \int_0^v b(v-x)f(x)\,dx$$

To maximize, we differentiate with respect to v and then set the derivative equal to 0.

$$0 = (b-c) + \int_v^\infty af(x)\,dx - \int_0^v bf(x)\,dx$$
$$0 = (b-c) + aP(X > v) - b(1 - P(X > v))$$
$$c = (a+b)P(X > v)$$

So the maximum profit is obtained when we pick v to satisfy

$$P(X > v) = c/(a+b)$$

If we take $a = b = c = 1$ then the objective becomes to maximize $-E|X-v|$ or to minimize $E|X-v|$, and the answer is that we pick v to satisfy $P(X > v) = 1/2$. This leads to the following property of the median

(2.5) Suppose $E|X| < \infty$. Let m be a median of X and c any other number. Then

$$E|X - c| \geq E|X - m|$$

We have proved this for the case in which X has a density. The key ideas for the proof in the discrete case can be found by solving

EXERCISE 2.4. In a small (and strange) town all the people live in four large apartment buildings named A, B, C, and D that are all on one road. It is 0.1 mile from A to B, 0.2 mile from B to C, and 0.7 mile from C to D. If we suppose 20% of the people live in A, 30% in B, 5% in C, and 45% in D, where should we locate the hospital to minimize the distance the ambulance will have to travel? Assume that all the people are equally likely to need an ambulance.

Before leaving this section we would like to note that (2.1) and (2.2) generalize in a straightforward way to functions of two or more random variables. If X and Y are discrete random variables then

$$(2.6) \qquad Er(X,Y) = \sum_x \sum_y r(x,y) P(X = x, Y = y)$$

If X and Y have joint density $f(x,y)$ then

$$(2.7) \qquad Er(X,Y) = \iint r(x,y) f(x,y)\, dy\, dx$$

EXERCISES

2.5. Suppose we roll two dice and let X and Y be the numbers they show. Find the expected value of (a) $X + Y$, (b) $\max\{X,Y\}$, (c) $\min\{X,Y\}$, (d) $|X - Y|$.

2.6. Five people play a game of "odd man out" to determine who will pay for the pizza they ordered. Each flips a coin. If only one person gets Heads (or Tails) while the other four get Tails (or Heads) then he is the odd man and has to pay. Otherwise they flip again. What is the expected number of tosses needed to determine who will pay?

2.7. Suppose $X = \text{Poisson}(\lambda)$. Find $E(1/(X+1))$.

2.8. Suppose X has a geometric distribution. Find the generating function $\gamma_X(z) = Ez^X$ and differentiate to find EX.

2.9. Suppose X has a uniform distribution on $\{0,1,\ldots,n-1\}$. Find the generating function $\gamma_X(z) = Ez^X$.

2.10. Suppose X has an exponential(λ) distribution. Find $E\sin X$.

2.11. Suppose X has density function $3x(2-x)/4$ for $0 < x < 2$ and 0 otherwise. Find EX^3.

2.12. Suppose X has the standard normal distribution. Use integration by parts to conclude that $EX^k = (k-1)EX^{k-2}$ and then conclude that for all integers n, $EX^{2n-1} = 0$ and $EX^{2n} = (2n-1)(2n-3)\cdots 3 \cdot 1$.

2.13. Suppose X has the standard normal distribution. Find Ee^{tX}.

2.14. If $\phi_X(t)$ is finite for $t \in (-t_0, t_0)$ then

$$\phi_X(t) = \sum_{k=0}^{\infty} \frac{t^k}{k!} \phi_X'(0) = \sum_{k=0}^{\infty} \frac{t^k}{k!} EX^k$$

Combine this with the formula in the previous problem to compute the moments of the normal distribution.

2.15. Suppose X has a uniform distribution on $(0,1)$. Compute Ee^{tX} and use the last problem to find all the moments EX^k. Of course the moments of the uniform distribution are easy to compute directly. (See Example 2.2.)

2.16. Suppose X has density function $1 - |x|$ for $-1 < x < 1$, 0 otherwise. Find the moment generating function for X.

2.17. Is $(3e^{4t} + e^{-2t})/4$ a moment generating function? If so, find the underlying distribution.

2.18. Is it possible for the moment generating function to be a polynomial $1 + \sum_{m=1}^{n} a_m t^m$ with $n \geq 1$ and $a_n \neq 0$?

2.19. $\kappa_X(t) = \ln \phi_X(t)$ is called the **cumulant generating function**. Show that $\kappa_X'(0) = EX$ and $\kappa_X''(0) = \mathrm{var}(X)$.

2.20. Suppose X and Y have joint density $x + y$ when $0 < x < 1$ and $0 < y < 1$, 0 otherwise. Compute EXY.

2.21. Two people agree to meet for a drink after work and each arrives independently at a time uniformly distributed between 5 p.m. and 6 p.m. What is the expected amount of time that the first person to arrive has to wait for the arrival of the second?

2.22. An archer shoots at a target. The displacement (X, Y) of her arrow from the center, measured in feet, has distribution $(1/(2\pi)e^{-(x^2+y^2)/2}$. Suppose the target consists of a circular bullseye and two circular rings, each having radius six inches, in which you get 5, 3, or 1 point. Find the archer's expected score from one arrow.

2.23. Suppose X and Y are independent and exponential(λ). Find $E|X - Y|$.

2.24. Suppose U and V are independent and uniform on $(0,1)$. Find $E(U-V)^2$.

2.25. Suppose $X > 0$. Is $E(1/X) = 1/EX$? Give a proof or a counterexample.

2.26. Suppose $f(0) = 0$ and $a_k = f(k) - f(k-1) \geq 0$ for $k \geq 1$. Imitate the proof of (1.3) to show that $Ef(X) = \sum_{k=1}^{\infty} a_k P(N \geq k)$. In particular $EN^2 = \sum_{k=1}^{\infty} (2k+1)P(N \geq k)$.

2.27. Suppose $X \geq 0$ has density f, $p > 0$ and $EX^p < \infty$. Use integration by parts to show that $EX^p = \int_0^\infty px^{p-1}P(X > x)\,dx$. To do this it is useful to note that

$$x^p P(X > x) \leq \int_x^\infty y^p f(y)\,dy \to 0 \qquad \text{as } x \to \infty$$

2.28. Show that $EX^p = \int_0^\infty px^{p-1}P(X > x)\,dx$ is true in the discrete case as well. To get started, notice that $P(X > x)$ is constant on $(k-1, k)$ so

$$\int_0^\infty px^{p-1}P(X > x)\,dx = \sum_{k=1}^\infty P(X \geq k) \int_{k-1}^k px^{p-1}\,dx$$

Then write $P(X \geq k) = \sum_{j=k}^\infty P(X = j)$ and interchange the order of summation.

2.29. The owner of a dress store buys dresses for \$50 and sells them for \$125. Any unsold dresses at the end of the season are sold to a second-hand store for \$25. Suppose to simplify things that the demand for dresses in one season is X, a random variable with density function $f(x)$. How many dresses should the store owner buy to maximize her profit?

2.30. A man sits in the lobby of a big office building selling newspapers. He buys papers for 15 cents and sells them for 25 cents but cannot return any papers that he does not sell. Suppose X people buy papers. Compute the difference between the profit from buying m papers and from buying $m-1$ papers and use this to determine the number of papers he should buy to maximize his profit.

2.31. Write a computer program to solve the last problem when X has a binomial distribution with $n = 120$ and $p = 1/3$.

Gambler's Ruin. Suppose a gambler starts with \$1 and plays a game in which he wins \$1 with probability p and loses \$1 with probability $1 - p$. Let f_n be the probability he first becomes broke at time n. Clearly, $f_0 = 0$ and $f_1 = 1 - p$. To compute the other values of f_k for $k \geq 2$ we notice that

$$f_k = p \sum_{j=0}^{k-1} f_j f_{k-1-j} \qquad \text{for } k \geq 2$$

since if he does not go broke at time 1 he has \$2 then, and in order to reach 0 from 2 in the remaining $k - 1$ steps, he must go down from 2 to 1 in j steps and then from 1 to 0 in the remaining $k - 1 - j$ steps. Multiplying both sides by z^k, summing, and recalling that $f_1 = 1 - p$, we have that the generating function of f, $\gamma(z) = \sum_k f_k z^k$ satisfies

$$\gamma(z) = (1-p)z + \sum_{k=2}^{\infty} p z^k \sum_{j=0}^{k-1} f_j f_{k-1-j}$$

$$= (1-p)z + pz \sum_{k=2}^{\infty} \sum_{j=0}^{k-1} z^j f_j \, z^{k-1-j} f_{k-1-j}$$

$$= (1-p)z + pz \sum_{j=0}^{\infty} \sum_{k=j+1}^{\infty} z^j f_j z^{k-(j+1)} f_{k-(j+1)}$$

$$= (1-p)z + pz \sum_{j=0}^{\infty} z^j f_j \sum_{m=0}^{\infty} z^m f_m$$

$$= (1-p)z + pz \, \gamma(z)^2$$

This is a quadratic equation $ax^2 + bx + c = 0$ with $x = \gamma(z)$, $a = pz$, $b = -1$, and $c = (1-p)z$ so its roots are

$$\frac{-b \pm \sqrt{b^2 - 4ac}}{2a} = \frac{1 \pm \sqrt{1 - 4p(1-p)z^2}}{2pz}$$

The answer we want is 0 at $z = 0$ so

$$\gamma(z) = \frac{1 - \sqrt{1 - 4p(1-p)z^2}}{2pz}$$

This formula was a little painful to derive but it contains a lot of interesting information. If we suppose $p > 1/2$ then the probability of ever hitting 0 is

$$\sum_{k=1}^{\infty} f_k = \gamma(1) = \frac{1 - \sqrt{1 - 4p(1-p)}}{2p} = (1-p)/p$$

since $1 - 4p + 4p^2 = (2p - 1)^2$.

If $p < 1/2$ then the expected time to hit 0, $\sum_{k=1}^{\infty} k f_k = \gamma'(1)$. To find this we observe that

$$\gamma'(z) = -\frac{1 - \sqrt{1 - 4p(1-p)z^2}}{2pz^2} - \frac{1}{2} \frac{(1 - 4p(1-p)z^2)^{-1/2}}{2pz}(-8p(1-p)z)$$

Setting $z = 1$ and observing that $\sqrt{1 - 4p + 4p^2} = (1 - 2p)$ (recall that $p < 1/2$ here) we have

$$\gamma'(1) = -\frac{1 - (1 - 2p)}{2p} + \frac{(1 - 2p)^{-1}}{4p} 8p(1-p)$$

$$= -1 + \frac{2(1-p)}{1 - 2p} = \frac{-(1 - 2p) + 2 - 2p}{1 - 2p} = \frac{1}{1 - 2p}$$

In hindsight, the last result is obvious: We expect to lose $(1 - 2p)$ dollars per play so the expected number of plays to lose one dollar is $1/(1 - 2p)$. Note that the last quantity approaches ∞ as $p \to 1/2$.

With some inspired guessing, we can even solve for the f_k exactly. Starting with Newton's binomial formula, (1.5) in Chapter 2, we find that

$$(1 - 4p(1-p)z^2)^{1/2} = \sum_{k=0}^{\infty} \binom{1/2}{k} (-4p(1-p)z^2)^k$$

Since the $k = 0$ term in the sum is 1,

$$1 - (1 - 4p(1-p)z^2)^{1/2} = -\sum_{k=1}^{\infty} \binom{1/2}{k} (-4p(1-p)z^2)^k$$

$$\frac{1 - (1 - 4p(1-p)z^2)^{1/2}}{2pz} = \sum_{k=1}^{\infty} \frac{(-1)^{k+1}}{2p} \binom{1/2}{k} (4p(1-p))^k z^{2k-1}$$

Since the right-hand side is $\gamma(z) = \sum_{k=1}^{\infty} f_{2k-1} z^{2k-1}$ (the even terms vanish since it is impossible to return to 0 with an even number of steps), we have

$$f_{2k-1} = \frac{(-1)^{k+1}}{2p} \binom{1/2}{k} (4p(1-p))^k$$

To get the formula in Example 1.9 we set $p = 1/2$ and observe that

$$\binom{1/2}{k} = \frac{(1/2)(-1/2)(-3/2) \cdots (-(2k-3)/2)}{k!}$$

so since $2p = 1$ and $4p(1-p) = 1$,

$$f_{2k-1} = (-1)^{k+1} \binom{1/2}{k} = (-1)^{k+1} 2^{-k} \frac{(-1)(-3) \cdots (-(2k-3))}{k!}$$

$$= (-1)^2 2^{-2k} \frac{2 \cdot 4 \cdots 2k}{k!} \frac{1 \cdot 3 \cdots (2k-3)}{k!}$$

$$= \frac{1}{2k-1} \binom{2k}{k} 2^{-2k}$$

The random variable N in Example 1.9 is the first time the number of Heads equals the number of Tails, so

$$P(N = 2k) = f_{2k-1} = \frac{1}{2k-1} \binom{2k}{k} 2^{-2k}$$

4.3. Properties of Expectation

In this section we will derive several properties of expected value. The first property is simple but very useful.

(3.1) $$E(aX + b) = aEX + b$$

PROOF: We will give the proof only for the continuous case. For the discrete case just replace the integrals by sums. Using (2.2),

$$Er(X) = \int r(x)f(x)\,dx$$

and then the facts that $\int g + h = \int g + \int h$ and $\int cg = c\int g$, we have

$$E(aX + b) = \int (ax + b)\,f(x)\,dx = \int axf(x)\,dx + \int bf(x)\,dx$$

$$= a\int xf(x)\,dx + b\int f(x)\,dx = aEX + b$$

since $\int x\,f(x)\,dx = EX$ and $\int f(x)\,dx = 1$ $\square$

Note that when $a = 0$, (3.1) says

$$E(b) = b$$

That is, the expected value of a number is just the number itself. Another trivial but useful special case of (3.1) is

$$E(X - EX) = EX - EX = 0$$

That is, if we subtract the mean we get a random variable with mean 0. Our third example of the use of (3.1) is

Example 3.1. Mean of the normal distribution. Suppose first that X has the standard normal density $f(x) = (2\pi)^{-1/2}e^{-x^2/2}$

$$EX = (2\pi)^{-1/2}\int x\,e^{-x^2/2}\,dx = (2\pi)^{-1/2}\left(-e^{-x^2/2}\right)\Big|_{-\infty}^{\infty} = 0$$

Now $Y = \sigma X + \mu$ is normal(μ, σ^2) so it follows from (3.1) that $EY = \sigma EX + \mu$, i.e., the mean is μ.

Our next goal is to compute the moment generating function for the normal. To do this it is useful to notice the following scaling property for the moment generating function:

(3.2)
$$\phi_{aX+b}(t) = e^{tb}\phi_X(ta)$$

PROOF: $Ee^{t(aX+b)} = E(e^{taX}e^{tb}) = e^{tb}E(e^{taX}) = e^{tb}\phi_X(ta)$ □

Example 3.2. Moment generating function of the normal distribution.
Suppose first that X has the standard normal distribution. In this case we compute Ee^{tX} by manipulating the integrand to look like the normal density with parameters $\mu = t$ and $\sigma^2 = 1$, which integrates to 1. We begin by observing that
$$-\frac{(x-t)^2}{2} = -\frac{x^2}{2} + tx - \frac{t^2}{2}$$

so
$$Ee^{tX} = \int e^{tx}\frac{1}{\sqrt{2\pi}}e^{-x^2/2}\,dx$$
$$= e^{t^2/2}\int \frac{1}{\sqrt{2\pi}}e^{-(x-t)^2/2}\,dx = e^{t^2/2}$$

Suppose now that $Y = \sigma X + \mu$ so that $Y = \text{normal}(\mu, \sigma^2)$. Using (3.2), we have
$$Ee^{tY} = e^{t\mu}\phi_X(t\sigma) = e^{t\mu}e^{(t\sigma)^2/2} = \exp(\mu t + \sigma^2 t^2/2)$$

EXERCISE 3.1. Suppose $X = \text{normal}(\mu, \sigma^2)$. Differentiate the moment generating function three times to find EX, EX^2, and EX^3.

Our next topic is sums of random variables.

(3.3) For any random variables X and Y,
$$E(X + Y) = EX + EY$$

PROOF: Using formula (2.7),
$$Er(X,Y) = \int\!\!\int r(x,y)f(x,y)\,dx\,dy$$

then the fact that $\int\!\!\int(g+h) = \int\!\!\int g + \int\!\!\int h$, we have
$$E(X+Y) = \int\!\!\int(x+y)f(x,y)\,dx\,dy$$
$$= \int\!\!\int xf(x,y)\,dx\,dy + \int\!\!\int yf(x,y)\,dx\,dy = EX + EY$$ □

As our first application of (3.3) we will show that bigger random variables have bigger expected values.

(3.4) If $X \geq Y$ then $EX \geq EY$.

PROOF: We begin with the special case: If $X \geq 0$ then $EX \geq 0$. To prove this in the continuous case we note that since $P(X < 0) = 0$,

$$EX = \int_0^\infty x f(x) \, dx \geq 0$$

since the integrand is nonnegative. To get from the special case to the general case we note that (3.3) and (3.1) imply

$$E(X - Y) = EX + E(-Y) = EX - EY$$

So if $X \geq Y$ then $X - Y \geq 0$ and it follows that $EX - EY = E(X - Y) \geq 0$. $\square$

From (3.3) and induction it follows that

(3.5) For any random variables $X_1, \ldots, X_n$,

$$E(X_1 + \cdots + X_n) = EX_1 + \cdots + EX_n$$

PROOF: (3.3) gives the result for $n = 2$. Applying (3.3) to $X = X_1 + \cdots + X_n$ and $Y = X_{n+1}$ we see that if the result is true for n then

$$\begin{aligned} E(X_1 + \cdots + X_{n+1}) &= E(X_1 + \cdots + X_n) + EX_{n+1} \\ &= EX_1 + \cdots + EX_n + EX_{n+1} \end{aligned}$$

and the result holds for $n + 1$. $\square$

(3.5) is quite useful in computing expected values. We begin with a simple computation that prepares for the three examples that follow it.

Example 3.3. Bernoulli distribution. Suppose X is 1 with probability p and 0 with probability $1 - p$.

$$EX = p \cdot 1 + (1 - p) \cdot 0 = p$$

Example 3.4. Binomial distribution. Consider a sequence of independent trials in which success has probability p. Let $X_i = 1$ if the ith trial results in a success and 0 otherwise. $P(X_i = 1) = p$ so $EX_i = p$. The total number of successes, $S_n = X_1 + \cdots + X_n$, so (3.5) implies

$$ES_n = EX_1 + \cdots + EX_n = np$$

This answer is quite reasonable: the expected number of successes on one trial is p, so the expected number of successes in n trials is np.

Example 3.5. Matching. Returning to Example 6.2 of Chapter 1, n men and n women at a dance are paired at random. Let $X_i = 1$ if the ith man dances with his wife and 0 otherwise. $P(X_i = 1) = 1/n$ so $EX_i = 1/n$. The number of men who dance with their wives, $D_n = X_1 + \cdots + X_n$, so (3.5) implies

$$ED_n = EX_1 + \cdots + EX_n = n(1/n) = 1$$

no matter how many people are involved!

Example 3.6. Hypergeometric distribution. Consider an urn with M red balls and N black balls. Draw n balls out of the urn and let X_i be 1 if the ith ball drawn is red, 0 otherwise. $P(X_i = 1) = M/(M+N)$ so $EX_i = M/(M+N)$. The total number of red balls drawn, $R_n = X_1 + \cdots + X_n$, so (3.5) implies

$$ER_n = EX_1 + \cdots + EX_n = nM/(M + N)$$

The random variables X_i used in the last three examples are a useful bookkeeping device. They are known as **indicator random variables**. If A is an event, we can define a random variable $X(\omega) = 1$ for $\omega \in A$ and $X(\omega) = 0$ for $\omega \in A^c$ that indicates whether or not A occurred. It is easy to see that $EX = P(A)$ but this is often a very useful observation.

Example 3.7. Inclusion-exclusion formula. Let $A_1, \ldots, A_n$ be events, and let $X_i = 1$ on A_i and $X_i = 0$ on A_i^c. Let $A = \cup_{i=1}^n A_i$ and $Y = 1$ on A, $Y = 0$ on A^c. Clearly, $Y = 1 - \prod_{i=1}^n (1 - X_i)$ since $Y = 1$ if and only if at least one $X_i = 1$. (Here $\prod_{i=1}^n z_i$ is short for $z_1 z_2 \cdots z_n$.) Expanding out the product,

$$Y = \sum_{i=1}^n X_i - \sum_{i<j} X_i X_j + \sum_{i<j<k} X_i X_j X_k - \cdots + (-1)^{n+1} \prod_{i=1}^n X_i$$

Taking expected values and using (3.5) we have

$$P(A) = \sum_{i=1}^n P(A_i) - \sum_{i<j} P(A_i \cap A_j) + \cdots + (-1)^{n+1} P(A_1 \cap \cdots \cap A_n)$$

since $X_i X_j = 1$ on $A_i \cap A_j$, 0 otherwise, etc. This is (6.2) from Chapter 1.

EXERCISE 3.2. Let B be the event that exactly one of the A_i occurs, and let $Z = 1$ on B, $Z = 0$ otherwise. Use the identity $Z = \sum_{i=1}^{n} X_i \prod_{j \neq i}(1 - X_j)$ to conclude that

$$P(B) = \sum_{i=1}^{n} P(A_i) - 2 \sum_{i<j} P(A_i \cap A_j) + \cdots + (-1)^{n+1} n \, P(A_1 \cap \cdots \cap A_n)$$

Turning from sums to products of random variables, we have

(3.6) If X and Y are independent then

$$EXY = EX \, EY$$

PROOF: Since X and Y are independent the joint density $f(x,y)$ is the product of the marginal densities $f_X(x)$ and $f_Y(y)$. Using this in our formula for $Er(X,Y)$ (that is, (2.7)) we have

$$EXY = \iint xy f_X(x) f_Y(y) \, dx \, dy = \int y f_Y(y) \int x f_X(x) \, dx \, dy$$
$$= \int y f_Y(y) EX \, dy = EX \, EY \qquad \qquad \square$$

From (3.6) and induction it follows that

(3.7) If $X_1, \ldots, X_n$ are independent,

$$E(X_1 \cdots X_n) = EX_1 \cdots EX_n$$

PROOF: (3.6) gives the result for $n = 2$. Applying (3.6) to $X = X_1 \cdots X_n$ and $Y = X_{n+1}$, which are independent by (5.8) in Chapter 3, we see that if the result is true for n then

$$E(X_1 \cdots X_{n+1}) = E(X_1 \cdots X_n) \cdot EX_{n+1} = EX_1 \cdots EX_n \cdot EX_{n+1}$$

and the result holds for $n + 1$. $\qquad \qquad \square$

(3.3) says that $E(X + Y) = EX + EY$ holds for ANY random variables. The next example shows that $EXY = EX \, EY$ does not hold in general.

Example 3.8. Suppose X and Y have joint distribution given by

X	$Y = 1$	0
1	0	.3
0	.5	.2

We have arranged things so that XY is always 0 so $EXY = 0$. On the other hand, $EX = P(X = 1) = 0.3$ and $EY = P(Y = 1) = 0.5$ so

$$EXY = 0 < 0.15 = EX\,EY$$

Our next example shows that we may have $EXY = EX\,EY$ without X and Y being independent.

Example 3.9. Suppose X and Y have joint distribution given by

X	$Y = -1$	0	1	
1	0	.25	0	.25
0	.25	.25	.25	.75
	.25	.5	.25	

Again we have arranged things so that XY is always 0, so $EXY = 0$. The symmetry of the marginal distribution for Y (or simple arithmetic) shows $EY = 0$, so we have $EXY = 0 = EX\,EY$. X and Y are not independent since $P(X = 1, Y = -1) = 0 < P(X = 1)P(Y = -1)$.

EXERCISE 3.3. Show that if X and Y take only the values 0 and 1 and $EXY = EX\,EY$ then X and Y are independent. By working a little harder you can show that the last result is valid whenever X and Y take only two values (that is, $P(X \in \{a, b\}) = 1$ and $P(Y \in \{c, d\}) = 1$).

Combining (3.7) with (5.7) from Chapter 3 gives

(3.8) Suppose $X_1, \ldots, X_n$ are independent and $r_1, \ldots, r_n$ are functions with $E|r_i(X_i)| < \infty$. Then

$$E(r_1(X_1) \cdots r_n(X_n)) = E(r_1(X_1)) \cdots E(r_n(X_n))$$

PROOF: (5.7) from Chapter 3 implies that $r_1(X_1), \ldots, r_n(X_n)$ are independent, so the desired conclusion follows from (3.7). □

Taking $r_i(x) = e^{tx}$ in (3.8), it follows that

(3.9) If $X_1, \ldots, X_n$ are independent

$$Ee^{t(X_1 + \cdots + X_n)} = Ee^{tX_1} \cdots Ee^{tX_n}$$

In words, the moment generating function of the sum of n independent random variables is the product of the individual moment generating functions. By taking $r(x) = z^x$ in (3.8) we see that the generating function of the sum of n independent random variables is the product of the individual generating functions.

Example 3.10. Binomial distribution. Consider a sequence of independent trials in which success has probability p. Let $X_i = 1$ if the ith trial results in a success and 0 otherwise. It is easy to see that

$$Ee^{tX_i} = (1 - p) + pe^t \qquad Ez^{X_i} = (1 - p) + pz$$

so using (3.9), the total number of successes $S_n = X_1 + \cdots + X_n$ has

$$Ee^{tS_n} = (1 - p + pe^t)^n \qquad Ez^{S_n} = (1 - p + pz)^n$$

EXERCISE 3.4. Differentiate the moment generating function to find ES_n and ES_n^2.

EXERCISE 3.5. Differentiate the generating function to find all the factorial moments of S_n, i.e., $E\{S_n(S_n - 1) \cdots (S_n - k + 1)\}$.

Once we introduce one more fact, (3.9) will allow us to quickly obtain all of the results we had to work for in Section 3.7.

(3.10) Suppose two m.g.f.'s $\phi_X(t)$, $\phi_Y(t)$ are finite and equal for $t \in (-t_0, t_0)$ with $t_0 > 0$; then X and Y have the same distribution.

Binomial. If $X = \text{binomial}(n, p)$ and $Y = \text{binomial}(m, p)$ are independent then $X + Y = \text{binomial}(n + m, p)$. Using the formula for the m.g.f. from Example 3.10,

$$\phi_{X+Y}(t) = \phi_X(t)\phi_Y(t) = \{(1 - p) + pe^t\}^n\{(1 - p) + pe^t\}^m$$
$$= \{(1 - p) + pe^t\}^{n+m}$$

Poisson. If $X = \text{Poisson}(\lambda)$ and $Y = \text{Poisson}(\mu)$ are independent then $X + Y = \text{Poisson}(\lambda + \mu)$. Using the formula for the m.g.f. from Example 2.4,

$$\phi_{X+Y}(t) = \phi_X(t)\phi_Y(t) = \exp(\lambda(e^t - 1))\exp(\mu(e^t - 1))$$
$$= \exp(\{\lambda + \mu\}(e^t - 1))$$

Normal. If $X = \text{normal}(\mu, a)$ and $Y = \text{normal}(\nu, b)$ are independent then $X + Y = \text{normal}(\mu + \nu, a + b)$. Using the formula for the m.g.f. from Example 3.2,

$$\phi_{X+Y}(t) = \phi_X(t)\phi_Y(t) = \exp(\mu t + at^2/2)\exp(\nu t + bt^2/2)$$
$$= \exp(\{\mu + \nu\}t + \{a + b\}t^2/2)$$

Gamma. If $X = \text{gamma}(m, \lambda)$, and $Y = \text{gamma}(n, \lambda)$ are independent then $X + Y = \text{gamma}(m + n, \lambda)$. Using the formula for the m.g.f. from Example 2.3,

$$\phi_{X+Y}(t) = \phi_X(t)\phi_Y(t) = \frac{\lambda^m}{(\lambda - t)^m} \cdot \frac{\lambda^n}{(\lambda - t)^n} = \frac{\lambda^{m+n}}{(\lambda - t)^{m+n}}$$

(3.10) is hard to prove but it is easy to show that

(3.11) If $\gamma_X(z) = \gamma_Y(z)$ for $z \in [0,1]$ then for all k, $P(X = k) = P(Y = k)$.

PROOF: $\gamma_X(z) = \sum_{m=0}^{\infty} z^m P(X = m)$ so $\gamma_X(0) = P(X = 0)$. (Recall our convention that $z^0 = 1$ even when $z = 0$.) Differentiating and discarding terms that are 0, we have

$$\gamma'(z) = \sum_{m=1}^{\infty} m z^{m-1} P(X = m)$$
$$\gamma'(0) = P(X = 1)$$
$$\gamma''(z) = \sum_{m=2}^{\infty} m(m-1) z^{m-2} P(X = m)$$
$$\gamma''(0) = 2P(X = 2)$$

From these two formulas the pattern should be clear. Using $\gamma^{(k)}$ to denote the kth derivative, we have

$$\gamma^{(k)}(t) = \sum_{m=k}^{\infty} m(m-1)\cdots(m-k+1) z^{m-k} P(X = m)$$
$$\gamma^{(k)}(0) = k! P(X = k)$$

If $\gamma_X(z)$ and $\gamma_Y(z)$ agree on $[0,1]$ (or even on $[0, \epsilon)$ for some $\epsilon > 0$) then their derivatives at 0 must agree and hence $P(X = k) = P(Y = k)$ for all k. $\square$

EXERCISES

3.6. A man plays roulette and bets $1 on black 19 times. He wins $1 with probability 18/38 and loses $1 with probability 20/38. What are his expected winnings?

3.7. Suppose we draw 13 cards out of a deck of 52. What is the expected value of the number of aces we get?

3.8. Suppose we pick 3 students at random from a class with 10 boys and 15 girls. Let X be the number of boys selected and Y be the number of girls selected. Find $E(X - Y)$.

3.9. Let N_k be the number of independent trials we need to get k successes when success has probability p. Find EN_k.

3.10. Let N_k be the number of independent trials we need to get k successes when success has probability p. Find Ez^{N_k} and differentiate to find EN_k.

3.11. Suppose that $X_1, \ldots, X_n > 0$ are independent and all have the same distribution. Find $E(\sum_{i=1}^{m} X_i / \sum_{j=1}^{n} X_j)$.

3.12. Twelve ducks fly overhead. Each of 6 hunters picks one duck at random to aim at and kills it with probability 0.6. (a) What is the mean number of ducks that are killed? (b) What is the expected number of hunters who hit the duck they aim at?

3.13. Suppose we put 7 balls randomly into 5 boxes. What is the expected number of empty boxes?

3.14. Suppose we draw 5 cards out of a deck of 52. What is the expected number of different suits in our hand? For example, if we draw $K\spadesuit$ $3\spadesuit$ $10\heartsuit$ $8\heartsuit$ $6\clubsuit$ there are three different suits in our hand.

3.15. Suppose Noah started with n pairs of animals on the ark and m of them died. If we suppose that fate chose the m animals at random, what is the expected number of complete pairs that are left?

3.16. Suppose we draw cards out of a deck without replacement. How many cards do we expect to draw out before we get (a) an Ace? (b) a spade?

3.17. A man and wife decide that they will keep having children until they have one of each sex. Ignoring the possibility of twins and supposing that each trial is independent and results in a boy or girl with probability 1/2, what is the expected value of the number of children they will have?

3.18. Suppose we flip a coin repeatedly and let $X_1, X_2, \ldots$ be the outcomes. If we stop at the first time N so that $X_N = T$ and $X_{N-1} = H$, find EN.

3.19. Suppose we roll a die repeatedly until we see each number at least once. Let R be the number of rolls required. Find ER.

3.20. Suppose $X_1, \ldots, X_n$ are independent and have a continuous distribution F. We say that a record occurs at time m if X_m is larger than $X_1, \ldots, X_{m-1}$. How many records do we expect to see?

3.21. Suppose you want to collect a set of n baseball cards. Assuming that you buy them one at a time and each time get a randomly chosen card, how many cards will you need to buy?

3.22. Consider the matching problem of Example 3.5. Let X_i be 1 if the ith man dances with his own wife and 0 otherwise. Let $D = X_1 + \cdots + X_n$ be the number of men that dance with their own wives. Use the fact that

$$D(D-1) \cdots (D-k+1) = k! \sum_{i_1 < i_2 \ldots < k} X_{i_1} \cdots X_{i_k}$$

to show that

$$E\{D(D-1) \cdots (D-k+1)\} = \begin{cases} 1 & \text{when } k \le n \\ 0 & \text{when } k > n \end{cases}$$

Note that the first n factorial moments agree with those of the Poisson(1) distribution.

3.23. **Jensen's inequality.** A twice differentiable function $g(x)$ is convex if $g''(x) \ge 0$ or equivalently if it always lies above its tangent line, that is,

$$g(x) \ge g(x_0) + g'(x_0)(x - x_0)$$

Use the last observation with $x_0 = EX$ and (3.4) to prove that $Eg(X) \ge g(EX)$.

Wald's equation. Let $X_1, X_2, \ldots$ be independent and have the same distribution with $E|X_i| < \infty$, and let $S_n = X_1 + \cdots + X_n$. (3.5) tells us that $ES_n = nEX_1$. We will now show that for random times N that "do not anticipate the future" $ES_N = ENEX_1$. Our first step is to define the random times we consider. To motivate the next definition, think of a gambler who wins X_i on the ith play and based on the outcomes of her bets, will stop playing at time N. N is said to be a **stopping time** if the event that she plays at time k, $\{N > k\}$, depends only on the outcomes $X_1, \ldots, X_k$. For example, N might be $\min\{n : S_n > 0\}$ the first time she is ahead. In this case $\{N > k\} = \{S_1 \le 0, \ldots, S_k \le 0\}$ and it is clear that $\{N > k\}$ depends only on $X_1, \ldots, X_k$.

(3.12) Let $X_1, X_2, \ldots$ be independent and have the same distribution with $E|X_i| < \infty$, and let $S_n = X_1 + \cdots + X_n$. If N is a stopping time with $EN < \infty$ then

$$ES_N = EN\, EX_1$$

The moral of this result for gamblers is that if $EX_i < 0$ then no matter what rule with $EN < \infty$ you use to stop playing, $ES_N < 0$.

PROOF: Let $1_{\{N>k\}}$ denote the random variable that is 1 on the event $\{N > k\}$ and 0 otherwise. The motivation for introducing this notation is to be able to write

$$S_N = \sum_{k=1}^{\infty} X_k 1_{\{N>k-1\}}$$

and then take expected values, using a generalization of (3.5) to get

$$ES_N = \sum_{k=1}^{\infty} E(X_k 1_{\{N>k-1\}})$$

Now the assumption that N is a stopping time implies that $1_{\{N>k-1\}}$ is a function $g_{k-1}(X_1, \ldots, X_{k-1})$, so (5.8) in Chapter 3 implies that X_k and $1_{\{N>k-1\}}$ are independent. Using (3.6) and the fact that all our random variables have the same distribution it follows that

$$E(X_k 1_{\{N>k-1\}}) = EX_k P(N > k - 1) = EX_1 P(N \geq k)$$

Inserting the last formula into our expression for ES_N we have

$$ES_N = EX_1 \sum_{k=1}^{\infty} P(N \geq k) = EX_1 EN$$

by (1.3).   □

Example 3.11. Gambler's Ruin. (3.12) can be used to give an easy proof of a result we derived at the end of Section 4.2. Suppose $P(X = 1) = p$, $P(X_i = -1) = 1 - p$, and let $N = \min\{n : S_n = -1\}$. Since

$$\{N > k\} = \{S_1 \geq 0, S_2 \geq 0 \ldots, S_k \geq 0\}$$

it is clear that N is a stopping time. If $EN < \infty$ then (3.15) implies

$$-1 = ES_N = EX_1 EN = (2p - 1)EN$$

so $EN = 1/(1 - 2p)$. What we have shown is that

$(\star)$ If $EN < \infty$ then $EN = 1/(1 - 2p)$. .

When $p > 1/2$, $(\star)$ says that if $EN < \infty$ then $EN < 0$, which is impossible since $N \geq 0$, so we must have $EN = \infty$ in that case. When $p = 1/2$, $(\star)$ says that if $EN < \infty$ then $EN = \infty$ so we must have $EN = \infty$ in that case as well. To show that $EN < \infty$ when $p < 1/2$, we consider $N \wedge n = \min\{N, n\}$. $N \wedge n \leq n$ so $E(N \wedge n) \leq n < \infty$, and applying (3.15) gives

$$-1 \leq ES_{N \wedge n} = (2p - 1)E(N \wedge n)$$

and dividing by $(2p - 1) < 0$ we have

$$1/(1 - 2p) \geq E(N \wedge n)$$

The last equality holds for all n so letting $n \to \infty$ we have $EN \leq 1/(1 - 2p) < \infty$ and we can conclude that $EN = 1/(1 - 2p)$ when $p < 1/2$.

4.4. Variance and Covariance

If $EX^2 < \infty$ then the **variance** of X is defined to be

$$\mathrm{var}(X) = E(X - EX)^2$$

As we will see in this section, the variance measures how spread-out the distribution of X is. To illustrate this concept we will consider some examples. But first, we need a formula that enables us to more easily compute $\mathrm{var}(X)$.

(4.1) $$\mathrm{var}(X) = EX^2 - (EX)^2$$

PROOF: Letting $\mu = EX$ to make the computations easier to see, we have

$$\mathrm{var}(X) = E(X - \mu)^2 = E\{X^2 - 2\mu X + \mu^2\} = EX^2 - 2\mu EX + \mu^2$$

by (3.5) and the facts that $E(-2\mu X) = -2\mu EX$, $E(\mu^2) = \mu^2$. Substituting $\mu = EX$ now gives (4.1). □

The reader should note that EX^2 means the expected value of X^2 and in the proof $E(X - \mu)^2$ means the expected value of $(X - \mu)^2$. When we want the square of the expected value we will write $(EX)^2$. This convention is designed to cut down on parentheses.

Example 4.1.   Uniform distribution on (a,b). Suppose X has density function $f(x) = 1/(b-a)$ for $a < x < b$ and 0 otherwise. In Example 1.6 we learned that $EX = (a+b)/2$.

$$EX^2 = \int_a^b x^2 \frac{1}{(b-a)} dx = \frac{x^3}{3(b-a)} \Big|_a^b = \frac{b^3 - a^3}{3(b-a)} = \frac{b^2 + ab + a^2}{3}$$

So

$$\mathrm{var}(X) = EX^2 - (EX)^2 = \frac{b^2 + ab + a^2}{3} - \frac{b^2 + 2ab + a^2}{4}$$

$$= \frac{b^2 - 2ab + a^2}{12} = \frac{(b-a)^2}{12}$$

Notice that the variance depends only on the length of the interval and is proportional to the square of the length.

Example 4.2.   Gamma distribution. Suppose X has density function $f(x) = \lambda^n x^{n-1} e^{-\lambda x}/(n-1)!$ for $x \geq 0$ and 0 otherwise. In Example 1.7 we learned that $EX = n/\lambda$. To compute EX^2 we manipulate the integrand to make it look like the gamma density with parameters $n+2$ and λ, which integrates to 1.

$$EX^2 = \int_0^\infty x^2 \frac{\lambda^n x^{n-1}}{(n-1)!} e^{-\lambda x} dx$$

$$= \frac{(n+1)!}{\lambda^2 (n-1)!} \int_0^\infty \frac{\lambda^{n+2} x^{n+1}}{(n+1)!} e^{-\lambda x} dx = \frac{(n+1)n}{\lambda^2}$$

Thus

$$\mathrm{var}(X) = EX^2 - (EX)^2 = \frac{(n+1)n}{\lambda^2} - \frac{n^2}{\lambda^2} = \frac{n}{\lambda^2}$$

When $n = 1$ this says that the exponential distribution with parameter λ has variance $1/\lambda^2$.

Example 4.3.   Geometric distribution. Suppose $P(N = n) = (1-p)^{n-1}p$ for $n = 1, 2, \ldots$ and 0 otherwise. To compute the variance we begin by observing that $\sum_{n=0}^\infty x^n = (1-x)^{-1}$. Differentiating this identity twice and noticing that the $n = 0$ term in the first derivative is 0 gives

$$\sum_{n=1}^\infty n x^{n-1} = (1-x)^{-2} \qquad \sum_{n=1}^\infty n(n-1)x^{n-2} = 2(1-x)^{-3}$$

Setting $x = 1 - p$ gives

$$\sum_{n=1}^\infty n(1-p)^{n-1} = p^{-2} \qquad \sum_{n=1}^\infty n(n-1)(1-p)^{n-2} = 2p^{-3}$$

Multiplying both sides by p in the first case and $p(1-p)$ in the second, we have

$$EN = \sum_{n=1}^{\infty} n(1-p)^{n-1}p = p^{-1}$$

$$E\{N(N-1)\} = \sum_{n=1}^{\infty} n(n-1)(1-p)^{n-1}p = 2p^{-2}(1-p)$$

From this it follows that

$$EN^2 = E\{N(N-1)\} + EN = \frac{2-2p}{p^2} + \frac{1}{p} = (2-p)/p^2$$

$$\text{var}(N) = EN^2 - (EN)^2 = \frac{2-p}{p^2} - \frac{1}{p^2} = (1-p)/p^2$$

Example 4.4. Poisson distribution. Suppose $P(X = k) = e^{-\lambda}\lambda^k/k!$ for $k = 0, 1, 2, \ldots$ and 0 otherwise. In Example 1.5 we learned that $EX = \lambda$. To compute EX^2 we begin by observing that the $k = 0$ and 1 terms make no contribution to the sum, so

$$E\{X(X-1)\} = \sum_{k=2}^{\infty} k(k-1)e^{-\lambda}\frac{\lambda^k}{k!} = \lambda^2 \sum_{k=2}^{\infty} e^{-\lambda}\frac{\lambda^{k-2}}{(k-2)!} = \lambda^2$$

since $\sum_{k=2}^{\infty} P(X = k-2) = 1$. From this it follows that

$$EX^2 = E\{X(X-1)\} + EX = \lambda^2 + \lambda$$
$$\text{var}(X) = EX^2 - (EX)^2 = \lambda^2 + \lambda - (\lambda)^2 = \lambda$$

Our first property of the variance will help explain the formulas we got for the uniform and gamma distributions.

(4.2) $$\text{var}(aX + b) = a^2\text{var}(X)$$

In words, the variance is not changed by adding a constant to X, but multiplying X by a multiplies the variance by a^2.

PROOF: $E(aX + b) = aEX + b$ so

$$\text{var}(aX + b) = E\{(aX + b) - E(aX + b)\}^2 = E\{(aX - aEX)^2\}$$
$$= E\{a^2(X - EX)^2\} = a^2 E\{(X - EX)^2\} = a^2\text{var}(X) \qquad \square$$

Taking $a = 0$ in (4.2), we see that

$$\text{var}(b) = 0$$

That is, a number is not random and hence has no variance. Conversely,

(4.3) If $\text{var}(X) = 0$ then $X = EX$ with probability one.

PROOF: Let $1_{\{|X-EX|>c\}}$ denote the random variable that is 1 on $\{|X-EX| > c\}$ and 0 otherwise. By considering two cases $|X - EX| > c$ and $|X - EX| \leq c$ one sees that

$$|X - EX|^2 \geq c^2 1_{\{|X-EX|>c\}}$$

Taking expected values, using (3.4) and the assumption that $\text{var}(X) = 0$ it follows that

$$0 = E|X - EX|^2 \geq c^2 P(|X - EX| > c)$$

The last inequality implies $P(|X - EX| > c) = 0$ for any $c > 0$, so

$$P(|X - EX| \leq 0) = 1$$

and since $|X - EX| \geq 0$ the desired result follows. $\qquad\qquad\square$

Another simple but useful fact is that if X has mean μ and variance σ^2 then, using (3.1) and (4.2), it follows that $Y = (X - \mu)/\sigma$ has

$$EY = \frac{1}{\sigma}E(X - \mu) = 0 \qquad \text{var}(Y) = \frac{1}{\sigma^2}\text{var}(X) = 1$$

In words, if we subtract the mean and divide by the square root of the variance then we get a random variable with mean 0 and variance 1.

Our third application of (4.2) is to check some of our variance computations. If X is uniform on $(0, 1)$ then $Y = (b - a)X + a$ is uniform on (a, b), so (4.2) implies $\text{var}(Y) = (b - a)^2\text{var}(X)$, which agrees with our computation that $\text{var}(Y) = (b - a)^2/12$. Likewise if $X = \text{gamma}(n, 1)$ then $Y = X/\lambda$ is $\text{gamma}(n, \lambda)$ (see Exercise 7.3 in Chapter 3), so (4.2) implies $\text{var}(Y) = \text{var}(X)/\lambda^2$ in agreement with our computation that $\text{var}(Y) = n/\lambda^2$. Our final example in this direction is to use (4.2) to compute the variance of the normal distribution.

Example 4.5. Normal distribution. Suppose first that X has the standard normal density $f(x) = (2\pi)^{-1/2}e^{-x^2/2}$. In Example 3.1 we learned that $EX =$

0. To compute $\text{var}(X) = EX^2$ we integrate by parts with $g(x) = (2\pi)^{-1/2}x$ and $h'(x) = xe^{-x^2/2}$ to get

$$\int x^2 (2\pi)^{-1/2} e^{-x^2/2}\, dx = (2\pi)^{-1/2} x(-e^{-x^2/2})\Big|_{-\infty}^{\infty}$$

$$+ \int (2\pi)^{-1/2} e^{-x^2/2}\, dx = 0 + 1$$

since $\int f(x)\, dx = 1$. Now $Y = \sigma X + \mu$ is normal(μ, σ^2) so it follows from (4.2) that $\text{var}(Y) = \sigma^2 \text{var}(X) = \sigma^2$.

The scaling relationship (4.2) shows that if X is measured in feet then the variance is measured in feet^2. This motivates the definition of the **standard deviation** $\sigma(X) = \sqrt{\text{var}(X)}$, which is measured in the same units as X and has a nicer scaling property.

(4.4) $\sigma(aX + b) = |a|\sigma(X)$

We get the absolute value here since $\sqrt{a^2} = |a|$. The standard deviation should be thought of as the size of the "typical deviation from the mean." To illustrate this, we will consider a concrete example.

Example 4.6. Roll one die and let X be the number that appears.

$$EX = 1 \cdot \frac{1}{6} + 2 \cdot \frac{1}{6} + 3 \cdot \frac{1}{6} + 4 \cdot \frac{1}{6} + 5 \cdot \frac{1}{6} + 6 \cdot \frac{1}{6} = \frac{21}{6}$$

$$EX^2 = 1 \cdot \frac{1}{6} + 4 \cdot \frac{1}{6} + 9 \cdot \frac{1}{6} + 16 \cdot \frac{1}{6} + 25 \cdot \frac{1}{6} + 36 \cdot \frac{1}{6} = \frac{91}{6}$$

so

$$\text{var}(X) = EX^2 - (EX)^2 = \frac{91 \cdot 6 - 441}{36} = \frac{105}{36} = 2.9166$$

and $\sigma(X) = \sqrt{\text{var}(X)} = 1.7078$. To explain our remark that $\sigma(X)$ gives the size of the "typical deviation from the mean," we note that the deviation from the mean

$$|X - \mu| = \begin{cases} 0.5 & \text{when } X = 3, 4 \\ 1.5 & \text{when } X = 2, 5 \\ 2.5 & \text{when } X = 1, 6 \end{cases}$$

so $E|X - \mu| = 1.5$. The standard deviation $\sigma(X) = \sqrt{E|X - \mu|^2}$ is a slightly less intuitive way of averaging the deviations $|X - \mu|$ but one that has nicer properties. (Most notably, (4.7) below.)

Our next topic is the variance of sums. To state our first result we need a definition. The **covariance** of X and Y is

$$\text{cov}(X,Y) = E\{(X - EX)(Y - EY)\}$$

Now $\text{cov}(X,X) = E\{(X - EX)^2\} = \text{var}(X)$, so repeating the proof of (4.1) we can rewrite the definition of the covariance in a form that is more convenient for computations:

(4.5)
$$\begin{aligned}
\text{cov}(X,Y) &= E\{XY - YEX - XEY + EXEY\} \\
&= EXY - EXEY
\end{aligned}$$

Having defined the covariance we are ready to prove

(4.6)
$$\text{var}\left(\sum_{i=1}^{n} X_i\right) = \sum_{i=1}^{n} \text{var}(X_i) + 2 \sum_{1 \le i < j \le n} \text{cov}(X_i, X_j)$$

PROOF: Let $\mu_i = EX_i$. (3.5) implies $E\left(\sum_{i=1}^{n} X_i\right) = \sum_{i=1}^{n} \mu_i$, so

$$\text{var}\left(\sum_{i=1}^{n} X_i\right) = E\left(\sum_{i=1}^{n} X_i - \sum_{i=1}^{n} \mu_i\right)^2 = E\left(\sum_{i=1}^{n} (X_i - \mu_i)\right)^2$$

To evaluate the square of the sum we note that

$$\left(\sum_{i=1}^{n} a_i\right)^2 = \sum_{i=1}^{n}\sum_{j=1}^{n} a_i a_j = \sum_{i=1}^{n} a_i^2 + 2 \sum_{1 \le i < j \le n} a_i a_j$$

Using the last identity with $a_i = X_i - \mu_i$, and the fact that the expected value of the sum is the sum of the expected values, we have

$$\text{var}\left(\sum_{i=1}^{n} X_i\right) = \sum_{i=1}^{n} E(X_i - \mu_i)^2 + 2 \sum_{1 \le i < j \le n} E\{(X_i - \mu_i)(X_j - \mu_j)\}$$

which is the desired formula. $\qquad\square$

To state our next result we need another definition. Random variables $X_1, \ldots, X_n$ are said to be **uncorrelated** if $\text{cov}(X_i, X_j) = 0$ whenever $i \ne j$. (4.5) implies that $\text{cov}(X_i, X_j) = 0$ is equivalent to $EX_i X_j = EX_i EX_j$, so (3.6) tells us that independent random variables are uncorrelated.

(4.7) If $X_1, \ldots, X_n$ are uncorrelated then

$$\text{var}(X_1 + \cdots + X_n) = \text{var}(X_1) + \cdots + \text{var}(X_n).$$

PROOF: Our assumption implies that the second sum in (4.6) vanishes.   □

(4.7) is not true in general. Indeed (4.6) shows that $\text{var}(X_1 + X_2) = \text{var}(X_1) + \text{var}(X_2)$ if and only if $\text{cov}(X_1, X_2) = 0$. For concrete counterexamples, let X be a random variable with $0 < \text{var}(X) < \infty$ and consider

(a) $X_1 = X$, $X_2 = -X$,

$$\text{var}(X_1 + X_2) = 0 \neq 2\,\text{var}(X) = \text{var}(X_1) + \text{var}(X_2)$$

(b) $X_1 = X_2 = X$,

$$\text{var}(X_1 + X_2) = \text{var}(2X) = 4\,\text{var}(X) \neq 2\,\text{var}(X) = \text{var}(X_1) + \text{var}(X_2)$$

(4.7), or more generally (4.6), is useful in computing variances. We begin with a simple computation that prepares for the three examples that follow it.

Example 4.7. Bernoulli distribution. Suppose $X = 1$ with probability p and 0 with probability $(1 - p)$. As we observed in Example 3.3, $EX = p$. To compute $\text{var}(X) = EX^2 - (EX)^2$ we note that

$$EX^2 = p \cdot 1^2 + (1 - p) \cdot 0^2 = p$$

so

$$\text{var}(X) = p - p^2 = p(1 - p)$$

Example 4.8. Binomial distribution. Consider a sequence of independent trials in which success has probability p. Let $X_i = 1$ if the ith trial results in a success and 0 otherwise. The total number of successes, $S_n = X_1 + \cdots + X_n$, has a binomial distribution with parameters n and p. (4.7) implies

$$\text{var}(S_n) = n\,\text{var}(X_1) = np(1 - p)$$

by the computation in the last example.

Example 4.9. Matching. Returning to Example 6.2 in Chapter 1, n men and n women at a dance are paired at random. Let $X_i = 1$ if the ith man dances with his wife and 0 otherwise. $D_n = X_1 + \cdots + X_n$ counts the number of men who dance with their own wives. $P(X_i = 1) = 1/n$ so, as we observed in Example 3.5, (3.5) implies $ED_n = 1$. To compute $\text{var}(D_n)$ using (4.6) we have to compute $\text{cov}(X_i, X_j)$ for $i < j$. Since the X_i only take the values 0 and 1, $X_i X_j$ will always be 0 or 1 and

$$E(X_i X_j) = P(X_i = 1, X_j = 1) = \frac{1}{n(n-1)}$$

by a calculation in the example cited. Plugging the last answer into (4.5),

$$\text{cov}(X_i, X_j) = EX_i X_j - EX_i\, EX_j$$
$$= \frac{1}{n(n-1)} - \frac{1}{n^2} = \frac{1}{n^2(n-1)}$$

Using this result in (4.6) and recalling the variance of the Bernoulli distribution with $p = 1/n$, we see that

$$\text{var}(D_n) = n\,\text{var}(X_1) + 2\binom{n}{2}\text{cov}(X_1, X_2)$$
$$= n\frac{1}{n}\frac{n-1}{n} + n(n-1)\frac{1}{n^2(n-1)} = \frac{n-1}{n} + \frac{1}{n} = 1$$

a formula that is valid for any $n \geq 2$. ($\text{var}(D_1) = 0$ since $D_1 = 1$ with probability one.)

Example 4.10. Hypergeometric distribution. Consider an urn with M red balls and N black balls. Draw n balls out of the urn and let $X_i = 1$ if the ith ball drawn was red, 0 otherwise. $R_n = X_1 + \cdots + X_n$ counts the number of red balls drawn. $P(X_i = 1) = M/(M + N)$ so as we observed in Example 3.6, (3.5) implies $ER_n = nM/(M + N)$. To compute $\text{var}(R_n)$ using (4.6) we have to compute $\text{cov}(X_i, X_j)$ for $i < j$. As in the last example, $X_i X_j$ can only take the values 0 and 1 so

$$EX_i X_j = P(X_i = 1, X_j = 1) = \frac{M(M-1)}{(M+N)(M+N-1)}$$

since if we suppose that the balls are numbered, symmetry implies that the numbers of the balls obtained on draws i and j take on the $(M+N)(M+N-1)$ possible values with equal probability and $M(M-1)$ of these possibilities correspond to the drawing of two red balls. Plugging the last formula into (4.5) we have

$$\text{cov}(X_i, X_j) = EX_i X_j - EX_i EX_j$$
$$= \frac{M(M-1)}{(M+N)(M+N-1)} - \frac{M}{M+N}\frac{M}{M+N}$$
$$= \frac{M}{M+N} \cdot \frac{(M-1)(M+N) - M(M+N-1)}{(M+N)(M+N-1)}$$

Now

$$(M - 1)(M + N) = M^2 - M + MN - N$$
$$M(M + N - 1) = M^2 + MN - M$$

so it follows that

$$\text{cov}(X_i, X_j) = \frac{-MN}{(M + N)^2(M + N - 1)}$$

Using this result in (4.6) and recalling the variance of the Bernoulli distribution with $p = M/(M + N)$, we see that

$$\text{var}(R_n) = n\text{var}(X_1) + 2\binom{n}{2}\text{cov}(X_1, X_2)$$

$$= n\frac{M}{M + N}\frac{N}{M + N} + n(n - 1)\frac{-MN}{(M + N)^2(M + N - 1)}$$

$$= \frac{nMN}{(M + N)^2}\frac{M + N - n}{M + N - 1}$$

$$= \frac{nMN}{(M + N)^2}\left(1 - \frac{n - 1}{M + N - 1}\right)$$

$$= np(1 - p)\frac{T - n}{T - 1}$$

where $p = M/(M + N)$ is the probability of a red ball on one draw and $T = M + N$ is the total number of balls in the urn. We have written the answer in this way to make it clear that (a) the variance of R_n is smaller than that of the binomial, $np(1 - p)$, with equality if and only if $n = 1$, and (b) the variance reaches a maximum when $n = T/2$ (or at the two integers nearest to this number when T is odd) and decreases to 0 when $n = T$.

Meaning of the sign of the covariance. In the matching example,

$$EX_1X_2 = P(X_1 = 1, X_2 = 1)$$

$$= P(X_1 = 1)P(X_2 = 1|X_1 = 1) = \frac{1}{n}\frac{1}{n - 1}$$

since if the first man picks his own wife, the second has probability $1/(n - 1)$ of picking his own wife. Since $P(X_2 = 1|X_1 = 1) > P(X_2 = 1)$, it follows that

$$\text{cov}(X_1, X_2) = EX_1X_2 - EX_1EX_2$$

$$= P(X_1 = 1)P(X_2 = 1|X_1 = 1) - P(X_1 = 1)P(X_2 = 1) > 0$$

In the hypergeometric example,

$$EX_1 X_2 = P(X_1 = 1, X_2 = 1)$$
$$= P(X_1 = 1)P(X_2 = 1|X_1 = 1) = \frac{M}{M+N}\frac{M-1}{M+N-1}$$

since if we get a red ball the first time, there are only $M - 1$ red balls in the $M + N - 1$ balls that remain. Since $P(X_2 = 1|X_1 = 1) < P(X_2 = 1)$, it follows that

$$\text{cov}(X_1, X_2) = EX_1 X_2 - EX_1 EX_2$$
$$= P(X_1 = 1)P(X_2 = 1|X_1 = 1) - P(X_1 = 1)P(X_2 = 1) < 0$$

In the matching example the covariance was positive because having $X_1 = 1$ made it more likely that $X_2 = 1$. In general, a positive covariance means that if X_1 is large then X_2 is more likely to be large. Concrete examples are $X_1 =$ a person's height and $X_2 =$ her weight, or X_1 and $X_2 =$ a student's scores on two exams in a class.

In the hypergeometric example the covariance was negative because having $X_1 = 1$ made it less likely that $X_2 = 1$. In general, a negative covariance means that if X_1 is large then X_2 is less likely to be large. Concrete examples are $X_1 =$ a person's weight and $X_2 =$ the number of hours per week she exercises, or $X_1 =$ the number of cigarettes an adult smokes per day and X_2 the amount of education she has $=$ the number of years of school completed.

EXERCISES

4.1. Suppose X has a uniform distribution on $\{1, 2, \ldots, n\}$. Find the mean and variance of X.

4.2. Find the mean and variance of the number of games in the World Series. Recall that it is won by the first team to win four games and assume that the outcomes are determined by flipping a coin.

4.3. Suppose two dice are rolled and let X be the larger of the two numbers that appear. Find the mean and the variance of X.

4.4. Suppose X has density function $3x^2$ for $0 < x < 1$, 0 otherwise. Find the mean and variance of X.

4.5. Suppose X has density function $(\rho - 1)x^{-\rho}$ for $x > 1$, 0 otherwise. Find EX^2 and the variance of X.

4.6. Suppose X has density function $(\lambda/2)e^{-\lambda|x|}$. Find the mean and variance of X.

4.7. Can we have a random variable with $EX = 3$ and $EX^2 = 8$?

4.8. Show that $E(X - c)^2$ is minimized by taking $c = EX$.

4.9. For a nonnegative random variable, the **coefficient of variation** is defined by $v(X) = \sigma(X)/EX$. Show that $v(X) \geq 1$.

4.10. Show that $\text{var}(X) = E\{X(X - 1)\} + EX - (EX)^2$ and use this to conclude that if γ_X is the generating function of X then $\text{var}(X) = \gamma_X''(1) + \gamma_X'(1) - (\gamma_X'(1))^2$.

4.11. Use the last exerecise to compute the variance of (a) the Poisson distribution, (b) the geometric distribution, and (c) the binomial distribution.

4.12. Suppose X and Y are independent with EX^2 and EY^2 finite. Find $\text{var}(3X - 4Y + 5)$.

4.13. In a class with 18 boys and 12 girls, boys have probability $1/3$ of knowing the answer and girls have probability $1/2$ of knowing the answer to a typical question the teacher asks. Assuming that whether or not the students know the answer are independent events, find the mean and variance of the number of students who know the answer.

4.14. Let N_k be the number of independent trials we need to get k successes when success has probability p. Find the variance of N_k.

4.15. Suppose we roll a die repeatedly until we see each number at least once and let R be the number of rolls required. Find the variance of R.

4.16. Suppose $X_1, \ldots, X_n$ are independent and have a continuous distribution F. We say that a record occurs at time m if X_m is larger than $X_1, \ldots, X_{m-1}$. What is the variance of the number of records?

4.17. Suppose you want to collect a set of n baseball cards. Assuming that you buy them one at a time and each time get a randomly chosen card, what is the variance of the number of cards you will need to buy?

4.18. Suppose we draw cards out of a deck without replacement. What is the variance of the number of cards we draw out before we get an ace?

4.19. Suppose $X_1, \ldots, X_n$ have $P(X_i = 1) = p$, $P(X_i = 0) = 1 - p$, and $P(X_i = 1, X_j = 1) = q$ whenever $i \neq j$. Show that

$$\text{var}(X_1 + \cdots + X_n) = np(1 - p) + n(n - 1)(q - p^2)$$

4.20. Suppose we put 7 balls randomly into 5 boxes. Find the variance of the number of empty boxes.

4.21. Suppose we draw 5 cards out of a deck of 52. What is the variance of the number of different suits in our hand? For example, if we draw $K\spadesuit$ $3\spadesuit$ $10\heartsuit$ $8\heartsuit$ $6\clubsuit$ there are three different suits in our hand.

4.22. Rank sums. Suppose $X_1, \ldots, X_m$ are independent random variables with a continuous distribution function F, and suppose $Y_1, \ldots, Y_n$ to be independent with a continuous distribution function G. Now order these $m + n$ random variables and let $Z_i = i$ if the ith smallest of the $m + n$ variables is an X and 0 otherwise. To test whether $F = G$, the Wilcoxon rank sum test looks at $R = Z_1 + \ldots + Z_n$ and decides that $F \neq G$ if R is too small or too large. Compute the mean and variance of R under the assumption that $F = G$. Note that the distribution of R does not depend on F (as long as F is continuous).

4.23. Suppose Noah started with n pairs of animals on the ark and m of them died. If we suppose that fate chose the m animals at random, what is the variance of the number of complete pairs that are left?

4.24. Suppose X takes on the values $-2, -1, 0, 1, 2$ with probability $1/5$ each, and let $Y = X^2$. (a) Find $\text{cov}(X, Y)$. (b) Are X and Y independent?

4.25. Suppose Z has a normal distribution with mean 0 and variance 1. Find $\text{cov}(Z, Z^2)$.

4.26. Suppose that X and Y have the same distribution, which has finite second moment. Show that $\text{cov}(X + Y, X - Y) = 0$. Notice that we are not assuming X and Y are independent.

4.27. Show that $\text{cov}(X, Y + Z) = \text{cov}(X, Y) + \text{cov}(X, Z)$ and use induction to conclude

$$\text{cov}\left(\sum_{i=1}^{m} X_i, \sum_{j=1}^{n} Y_j\right) = \sum_{i=1}^{m}\sum_{j=1}^{n} \text{cov}(X_i, Y_j)$$

4.28. Suppose we perform n independent experiments in which the disjoint events A and B have probabilities a and b. Let M be the number of times A occurs and N the number of times B occurs. Use the formula in the last exercise to show $\text{cov}(M, N) = -nab$.

4.29. Use (4.6) to derive the last result.

4.30. Suppose we draw n balls out of an urn that contains K red, L black, and M balls of other colors. Let U be the number of red balls and V be the number of black balls obtained. Find $\text{cov}(U, V)$.

*4.5. Correlation

If X and Y are random variables with standard deviations $\sigma(X)$ and $\sigma(Y)$ that, are positive we define the **correlation** to be

$$\rho(X,Y) = \frac{\operatorname{cov}(X,Y)}{\sigma(X)\sigma(Y)} = E\left\{ \frac{(X - EX)}{\sigma(X)} \cdot \frac{(Y - EY)}{\sigma(Y)} \right\}$$

In words, we subtract the mean and divide by the standard deviation of each variable to make their means 0 and variances 1, then we take the expected value of the product. (If $\sigma(X) = 0$ then X is a constant and is independent of Y so we set $\rho(X,Y) = 0$ when $\sigma(X)$ or $\sigma(Y) = 0$.)

An immediate consequence of the fact that we subtract the mean and divide by the standard deviation is that the correlation of two random variables is not affected by the units in which they are measured. For instance if X is height and Y is weight then it does not matter if we use inches or centimeters, and pounds or kilograms.

(5.1) If $Y_1 = a_1 X_1 + b_1$ and $Y_2 = a_2 X_2 + b_2$ then

$$\rho(Y_1, Y_2) = \frac{a_1}{|a_1|} \frac{a_2}{|a_2|} \rho(X_1, X_2)$$

PROOF: $EY_i = a_i EX_i + b_i$, so

$$Y_i - EY_i = (a_i X_i + b_i) - (a_i EX_i + b_i) = a_i(X_i - EX_i)$$

and hence

$$E\{(Y_1 - EY_1)(Y_2 - EY_2)\} = a_1 a_2 E\{(X_1 - EX_1)(X_2 - EX_2)\}$$

(4.4) implies $\sigma(Y_i) = |a_i|\sigma(X_i)$. So dividing both sides by

$$\sigma(Y_1)\sigma(Y_2) = |a_1||a_2|\sigma(X_1)\sigma(X_2)$$

we have the indicated result.   □

EXERCISE 5.1. Show that $\operatorname{cov}(a_1 X_1 + b_1, a_2 X_2 + b_2) = a_1 a_2 \operatorname{cov}(X_1, X_2)$.

When a_1 and a_2 are positive, which is the case when converting height from inches to centimeters, or temperatures from Fahrenheit to Centigrade, $a_1/|a_1| = a_2/|a_2| = 1$ and $\rho(Y_1, Y_2) = \rho(X_1, X_2)$. Taking $a_1 = 1$, $a_2 = -1$, and $b_1 = b_2 = 0$, we have

(5.2) $$\rho(X, -Y) = -\rho(X, Y)$$

Our next goal is to prove

(5.3) $-1 \le \rho(X,Y) \le 1$

with $|\rho(X,Y)| = 1$ if and only if $Y = aX + b$.

PROOF: We begin by observing that

$$\rho(X,X) = E\left\{\left(\frac{(X-EX)}{\sigma(X)}\right)^2\right\} = \frac{E(X-EX)^2}{\text{var}(X)} = 1$$

If $Y = aX + b$ then (5.1) implies

$$\rho(X,Y) = \frac{a}{|a|}\rho(X,X)$$

so $|\rho(X,Y)| = 1$ in this case. To prove that $\rho(X,Y)$ always lies between 1 and -1 we use the

(5.4) Cauchy-Schwarz inequality.

$$(EUV)^2 \le EU^2\, EV^2$$

PROOF OF (5.4): Clearly,

$$0 \le E(\theta U + V)^2 = \theta^2 EU^2 + 2\theta EUV + EV^2$$

Now if $a\theta^2 + b\theta + c \ge 0$ for all θ then we must have $b^2 - 4ac \le 0$. (Recall that the roots of the quadratic are $(-b \pm \sqrt{b^2 - 4ac})/2a$, so if $b^2 - 4ac > 0$ there are two real roots and the quadratic must be negative somewhere.) In our case $b = 2EUV$, $a = EU^2$, and $c = EV^2$ so $b^2 - 4ac \le 0$ means $4(EUV)^2 - 4EU^2EV^2 \le 0$, which, after a little rearranging, is the desired conclusion. $\square$

Letting $U = X - EX$ and $V = Y - EY$, (5.4) implies

$$(\text{cov}(X,Y))^2 = (EUV)^2 \le EU^2\, EV^2 = \text{var}(X)\text{var}(Y)$$

Taking square roots, we have $|\text{cov}(X,Y)| \le \sigma(X)\sigma(Y)$, so

$$|\rho(X,Y)| \le 1$$

The last detail is to show that if $|\rho(X,Y)| = 1$ then $Y = aX + b$. To do this, we begin by investigating the consequences of equality in (5.4). Suppose $EUV = \sqrt{EU^2 EV^2}$ and $EU^2 > 0$. Letting $\theta = \sqrt{EV^2/EU^2}$ we have

$$E(\theta U - V)^2 = \theta^2 EU^2 - 2\theta EUV + EV^2$$

$$= \frac{EV^2}{EU^2} EU^2 - 2\sqrt{EV^2/EU^2}\sqrt{EU^2 EV^2} + EV^2$$

$$= EV^2 - 2EV^2 + EV^2 = 0$$

so $E(\theta U - V)^2 = 0$ and it follows that $V = \theta U$ with probability one. If instead, $EUV = -\sqrt{EU^2 EV^2}$, we let $\theta = -\sqrt{EV^2/EU^2}$ and we get the same conclusion. At this point we have shown that if equality holds in (5.4) then U is θ times V. Going back to the proof of $-1 \le \rho(X,Y) \le 1$ we see that if $|\rho(X,Y)| = 1$ then there is equality when we apply (5.4) to $U = X - EX$ and $V = Y - EY$, so $V = Y - EY$ is θ times $U = X - EX$, and

$$Y = EY + \theta(X - EX) = aX + b \qquad \square$$

Having indulged in a lot of theory, we will now pause to consider a concrete example.

Example 5.1.

X	Y=0	1	2	
2	.2	.1	0	.3
1	.1	.2	.1	.4
0	0	.1	.2	.3
	.3	.4	.3	

From the marginal distributions it is easy to see that $EX = EY = 1$, $EX^2 = EY^2 = (0.3)0 + (0.4)1 + (0.3)4 = 1.6$, and $\text{var}(X) = EX^2 - (EX)^2 = 0.6 = \text{var}(Y)$. To compute EXY we note that

$$P(XY = 0) = P(X = 0 \text{ or } Y = 0) = 0.2 + 0.1 + 0.1 + 0.2 = 0.6$$
$$P(XY = 1) = P((X,Y) = (1,1)) = 0.2,$$
$$P(XY = 2) = P((X,Y) = (1,2) \text{ or } (2,1)) = 0.1 + 0.1 = 0.2$$

So $EXY = 0.2(1) + 0.2(2) = 0.6$ and

$$\text{cov}(X,Y) = EXY - EX\,EY = 0.6 - 1 \cdot 1 = -0.4$$

$$\rho(X,Y) = \frac{\text{cov}(X,Y)}{\sigma(X)\sigma(Y)} = \frac{-0.4}{0.6} = -\frac{2}{3}$$

To get a feeling for what $\rho(X,Y) = -2/3$ says about the relationship between X and Y we will now show how to construct random variables X_1 and X_2 with $EX_i = \mu_i$, $\mathrm{var}(X_i) = \sigma_i^2$, and $\rho(X_1, X_2) = \rho$.

Example 5.2. Let Z_1 and Z_2 be independent with mean 0 and variance 1. Let $Y_1 = Z_1$ and $Y_2 = \rho Z_1 + \sqrt{1 - \rho^2} Z_2$. Clearly $EY_1 = 0$, $\mathrm{var}(Y_1) = 1$, and $EY_2 = 0$. To compute $\mathrm{var}(Y_2) = EY_2^2$ we observe that

$$EY_2^2 = \rho^2 EZ_1^2 + 2\rho\sqrt{1 - \rho^2} EZ_1 Z_2 + (1 - \rho^2) EZ_2^2$$
$$= \rho^2 + 0 + (1 - \rho^2) = 1$$

since $EZ_1^2 = EZ_2^2 = 1$ and (3.6) implies $EZ_1 Z_2 = EZ_1 EZ_2 = 0$. To compute the correlation, we note that the means are 0 and the variances are 1, so

$$\rho(Y_1, Y_2) = EY_1 Y_2 = \rho EZ_1^2 + \sqrt{1 - \rho^2} EZ_1 Z_2 = \rho$$

since $EZ_1^2 = 1$ and $EZ_1 Z_2 = 0$. To construct the X_i now we let $X_i = \mu_i + \sigma_i Y_i$ so that $EX_i = \mu_i$ and $\mathrm{var}(X_i) = \sigma_i^2$, and (5.1) implies that $\rho(X_1, X_2) = \rho(Y_1, Y_2) = \rho$.

Combining the formulas for the X_i and Y_i we have

(5.5)
$$X_1 = \mu_1 + \sigma_1 Z_1$$
$$X_2 = \mu_2 + \sigma_2 \rho Z_1 + \sigma_2 \sqrt{1 - \rho^2} Z_2$$

The first equation says $Z_1 = (X_1 - \mu_1)/\sigma_1$. Plugging this into the second equation gives

(5.6)
$$X_2 = \mu_2 + \rho \frac{\sigma_2}{\sigma_1}(X_1 - \mu_1) + \sigma_2 \sqrt{1 - \rho^2} Z_2$$

Note that in this case X_2 is a linear function of X_1 plus a multiple of Z_2, which is independent of X_1. While this is just an example, the next set of computations will show that this is almost the general picture. The "almost" here refers to the fact that in general X_1 and Z_2 are not independent but are only uncorrelated.

Example 5.3. Linear prediction. The problem here is to pick a and b to minimize $E(X_2 - (aX_1 + b))^2$. For example, we might want to find the best linear function of a person's height X_1 to predict her weight X_2, where "best" means it minimizes the mean square error, $E(X_2 - (aX_1 + b))^2$. Expanding out the square and then using the facts that $E\left(\sum_i Y_i\right) = \sum_i EY_i$ and $E(cY) = cEY$, we have

$$E(X_2 - (aX_1 + b))^2 = E\{X_2^2 - 2X_2(aX_1 + b) + (aX_1 + b)^2\}$$
$$= E\{X_2^2 - 2aX_1 X_2 - 2bX_2 + a^2 X_1^2 + 2abX_1 + b^2\}$$
$$= EX_2^2 - 2aEX_1 X_2 - 2bEX_2 + a^2 EX_1^2 + 2abEX_1 + b^2$$

Taking partial derivatives with respect to a and b and setting them equal to 0,

(⋆)
$$0 = -2EX_1X_2 + 2aEX_1^2 + 2bEX_1$$
$$0 = -2EX_2 + 2aEX_1 + 2b$$

Multiplying the second equation by $-EX_1$ and adding it to the first, we have

$$0 = -2EX_1X_2 + 2EX_1EX_2 + 2aEX_1^2 - 2a(EX_1)^2$$

Dividing by 2 and rearranging, we have

$$EX_1X_2 - EX_1EX_2 = a\{EX_1^2 - (EX_1)^2\}$$

$$a = \frac{\text{cov}(X_1, X_2)}{\text{var}(X_1)} = \frac{\rho\sigma_1\sigma_2}{\sigma_1^2} = \rho\frac{\sigma_2}{\sigma_1}$$

where $\rho = \rho(X_1, X_2)$ and σ_i is the standard deviation of X_i. To find b we notice that the second equation in (⋆) implies $b = EX_2 - aEX_1 = \mu_2 - a\mu_1$, so the best straight line fit is

$$b + aX_1 = \mu_2 + a(X_1 - \mu_1) = \mu_2 + \rho\frac{\sigma_2}{\sigma_1}(X_1 - \mu_1)$$

which matches the first part of (5.6). The mean square error of the prediction, $aX_1 + b$, is given by

$$E\left(X_2 - \mu_2 - \rho\frac{\sigma_2}{\sigma_1}(X_1 - \mu_1)\right)^2$$

$$= E(X_2 - \mu_2)^2 + \rho^2\frac{\sigma_2^2}{\sigma_1^2}E(X_1 - \mu_1)^2 - 2\rho\frac{\sigma_2}{\sigma_1}E\{(X_2 - \mu_2)(X_1 - \mu_1)\}$$

$$= \sigma_2^2 + \rho^2\sigma_2^2 - 2\rho^2\sigma_2^2 = \sigma_2^2(1 - \rho^2)$$

since $\text{cov}(X_1, X_2) = \rho\sigma_1\sigma_2$. This computation tells us that $\sigma_2^2(1 - \rho^2)$ is the amount of variance that is not accounted for by the straight line fit. Notice that this is the variance of the last term in (5.6).

The last step in justifying the remark we made just before this example is to show that the "error" $\Delta = X_2 - (aX_1 + b)$ has $\text{cov}(X_1, \Delta) = 0$. To prove this, we use Exercises 4.27 and 5.1 to conclude that

$$\text{cov}(X_1, X_2 - (aX_1 + b)) = \text{cov}(X_1, X_2) + \text{cov}(X_1, -(aX_1 + b))$$
$$= \text{cov}(X_1, X_2) - a\,\text{cov}(X_1, X_1)$$
$$= \rho\sigma_1\sigma_2 - \rho\frac{\sigma_2}{\sigma_1}\sigma_1^2 = 0$$

Example 5.4. In our concrete example

$$a = \rho \frac{\sigma_2}{\sigma_1} = -2/3 \qquad b = \mu_2 - a\mu_1 = 1 - \frac{-2}{3} \cdot 1 = 5/3$$

so the best linear predictor is $(-2/3)X_1 + (5/3)$. The arithmetic is somewhat simpler and the answer more transparent if we use $b = \mu_2 - a\mu_1$ to write the best linear predictor as

$$\mu_2 + a(X_1 - \mu_1) = 1 - (2/3)(X_1 - 1)$$

This form makes it clear, for instance, that (μ_1, μ_2) is always on the line. No matter how you write the line, the error $\Delta = X_2 - (aX_1 + b)$ has

$$\frac{\text{var}(\Delta)}{\sigma_2^2} = (1 - \rho^2) = 1 - \left(\frac{2}{3}\right)^2 = 5/9$$

so the straight line fit only "explains" 4/9 of the variance.

Example 5.5. For a second concrete example we will now consider the situation in which X and Y have joint density

$$f(x, y) = \begin{cases} 8xy & \text{when } 0 < y < x < 1 \\ 0 & \text{otherwise} \end{cases}$$

To plug into the formulas above we will take $X_1 = X$ and $X_2 = Y$. To cut down on the number of computations required to determine the best linear predictor $\mu_2 + a(X_1 - \mu_1)$ we note that $\rho = \text{cov}(X_1, X_2)/\sigma_1\sigma_2$ so

$$a = \rho \frac{\sigma_2}{\sigma_1} = \frac{\text{cov}(X, Y)}{\text{var}(X)}$$

To simplify the calculus we observe

$$\int_0^1 \int_0^x x^m y^n \, dy \, dx = \int_0^1 \frac{x^{m+n+1}}{n+1} \, dx = \frac{1}{(n+1)(m+n+2)}$$

Plugging into the last formula, we find

$m = 2$	$n = 1$	EX	$= 8/(2 \cdot 5)$	$= 4/5$
$m = 1$	$n = 2$	EY	$= 8/(3 \cdot 5)$	$= 8/15$
$m = 3$	$n = 1$	EX^2	$= 8/(2 \cdot 6)$	$= 2/3$
$m = 2$	$n = 2$	EXY	$= 8/(3 \cdot 6)$	$= 4/9$
$m = 1$	$n = 3$	EY^2	$= 8/(4 \cdot 6)$	$= 1/3$

which gives us

$$\text{var}(X) = 2/3 - 16/25 = 2/75 = 6/225$$
$$\text{cov}(X,Y) = 4/9 - 32/75 = 4/225$$
$$\text{var}(Y) = 1/3 - 64/225 = 11/225$$

Thus $a = \text{cov}(X,Y)/\text{var}(X) = 2/3$ and the best linear predictor is

$$\mu_2 + a(X_1 - \mu_1) = 8/15 + (2/3)(X - 4/5) = 2X/3$$

The correlation

$$\rho = \frac{\text{cov}(X,Y)}{\sigma_X \sigma_Y} = 4/\sqrt{66} = 0.4924$$

and the variance not explained by the straight line fit is

$$(1 - \rho^2)\sigma_2^2 = \frac{50}{66} \cdot \frac{11}{225} = 1/27$$

Example 5.6. The bivariate normal distribution. In this example we will use (5.5) to "derive" the bivariate normal distribution. The resulting formula is somewhat awful but this is one of our points: It is much easier to understand the bivariate normal by looking at (5.5) and (5.6) than at the formula for its joint density.

Suppose Z_1 and Z_2 are independent standard normals and let

$$Y_1 = \sigma_1 Z_1 \qquad Y_2 = \sigma_2 \rho Z_1 + \sigma_2 \sqrt{1 - \rho^2} Z_2$$

By results in Example 5.2, $EY_i = 0$, $\text{var}(Y_i) = \sigma_i^2$, and $\rho(Y_1, Y_2) = \rho$. Since the $Y_i = r_i(Z_1, Z_2)$ we can compute the joint density of (Y_1, Y_2) using the results of Section 3.6. The first step in doing this is to find the inverse functions. Setting

$$y_1 = \sigma_1 z_1 \qquad y_2 = \sigma_2 \rho z_1 + \sigma_2 \sqrt{1 - \rho^2} z_2$$

and solving for z_1 and z_2 gives

$$s_1(y_1, y_2) = y_1/\sigma_1$$
$$s_2(y_1, y_2) = \frac{y_2 - \sigma_2 \rho z_1}{\sigma_2 \sqrt{1 - \rho^2}} = \frac{1}{\sqrt{1 - \rho^2}}\left(\frac{y_2}{\sigma_2} - \frac{\rho y_1}{\sigma_1}\right)$$

Calculating partial derivatives we find

$$D_{11} = \partial s_1/\partial y_1 = 1/\sigma_1 \qquad\qquad D_{12} = \partial s_1/\partial y_2 = 0$$

$$D_{21} = \partial s_2/\partial y_1 = -\rho/\sigma_1\sqrt{1-\rho^2} \quad D_{22} = \partial s_2/\partial y_2 = 1/\sigma_2\sqrt{1-\rho^2}$$

so the determinant

$$J = D_{11}D_{22} - D_{12}D_{21} = 1/\sigma_1\sigma_2\sqrt{1-\rho^2}$$

Recalling that

$$f_Z(z_1, z_2) = \frac{1}{2\pi}\exp(-(z_1^2 + z_2^2)/2)$$

we now have all the ingredients for using formula (6.1) in Chapter 3.

$$f_Y(y) = |J|f_Z(s(y)) = \frac{1}{\sigma_1\sigma_2\sqrt{1-\rho^2}} \cdot \frac{1}{2\pi}\exp\left(-\frac{y_1^2}{2\sigma_1^2}\right)$$

$$\cdot \exp\left(-\frac{1}{2(1-\rho^2)}\left\{\frac{\rho^2 y_1^2}{\sigma_1^2} - \frac{2\rho y_1 y_2}{\sigma_1\sigma_2} + \frac{y_2^2}{\sigma_2^2}\right\}\right)$$

$$= \frac{1}{2\pi\sigma_1\sigma_2\sqrt{(1-\rho^2)}} \cdot \exp\left(-\frac{1}{2(1-\rho^2)}\left\{\frac{y_1^2}{\sigma_1^2} - \frac{2\rho y_1 y_2}{\sigma_1\sigma_2} + \frac{y_2^2}{\sigma_2^2}\right\}\right)$$

The last formula is for the case $\mu_1 = \mu_2 = 0$ but introducing non-zero means is easy now. Let $X_i = Y_i + \mu_i$ and observe that
(5.7)

$$f_{X_1, X_2}(x_1, x_2) = f_{Y_1, Y_2}(x_1 - \mu_1, x_2 - \mu_2) = \frac{1}{2\pi\sigma_1\sigma_2\sqrt{(1-\rho^2)}}$$

$$\exp\left(-\frac{1}{2(1-\rho^2)}\left\{\frac{(x_1 - \mu_1)^2}{\sigma_1^2} - \frac{2\rho(x_1 - \mu_1)(x_2 - \mu_2)}{\sigma_1\sigma_2} + \frac{(x_2 - \mu_2)^2}{\sigma_2^2}\right\}\right)$$

This is the **bivariate normal density**. As we will see in the next section, it is much easier to think of X_1 and X_2 as being

(5.8)
$$X_1 = \mu_1 + \sigma_1 Z_1$$
$$X_2 = \mu_2 + \sigma_2\rho Z_1 + \sigma_2\sqrt{1-\rho^2}Z_2$$

where Z_1 and Z_2 are independent standard normals.

From (5.7) or (5.8) one can see that if $\rho = 0$ then X_1 and X_2 are independent. In words, uncorrelated bivariate normal random variables are independent.

EXERCISES

5.2. Suppose X and Y have joint density $f(x, y) = 2$ for $0 < y < x < 1$, 0 otherwise. Find (a) $\rho(X, Y)$, (b) the best linear predictor $aX + b$ of Y.

5.3. Suppose X and Y have joint density $f(x,y) = 24xy$ when $x > 0$, $y > 0$, and $x + y < 1$, 0 otherwise. Find (a) $\rho(X,Y)$, (b) the best linear predictor $aX + b$ of Y. To cope with the calculus it is useful to note that

$$\int_0^1 \int_0^{1-x} x^m y^n \, dy \, dx = \int_0^1 \frac{x^m (1-x)^{n+1}}{n+1} \, dx = \frac{m! \, n!}{(m+n+2)!}$$

with the second equality the result of repeated integration by parts.

5.4. Suppose X and Y have joint density $f(x,y) = x + y$ when $0 < x < 1$ and $0 < y < 1$, 0 otherwise. Find (a) $\rho(X,Y)$, (b) the best linear predictor $aX + b$ of Y.

5.5. Suppose X and Y have joint density $f(x,y) = e^{-y}$ for $0 < x < y$, 0 otherwise. Find (a) $\rho(X,Y)$, (b) the best linear predictor $aX + b$ of Y.

5.6. Let Z_1 and Z_2 be independent with means m_i and variances v_i^2. Let $X = Z_1$ and $Y = Z_1 + Z_2$. Find (a) $\rho(X,Y)$, (b) the best linear predictor $aX + b$ of Y.

5.7. Suppose $X_1, \ldots, X_n$ are independent with $EX_i = \mu_i$ and $\text{var}(X_i) = \sigma^2$. Let $S_k = X_1 + \cdots + X_k$. Find $\rho(S_k, S_n)$.

5.8. **Time series model of air pollution.** Suppose X_n represents the amount of pollution put into the air on day n, suppose that a fraction $1-\theta$ breaks down each day, and let

$$Y_n = \sum_{m=0}^{\infty} \theta^m X_{n-m}$$

be the amount that is present at time n. Assuming that the X_n are independent with mean μ and variance σ^2, compute (a) EY_i, (b) $\text{var}(Y_i)$, and (c) $\rho(Y_0, Y_n)$ for $n > 0$. For the last part, it is useful to note that $Y_n = \theta^n Y_0 + Z$ where $Z = \theta^{n-1} X_1 + \cdots + X_n$ is independent of Y_0.

5.9. Suppose that U is uniform on $(0,1)$, $X = U^2$, and $Y = U^3$. (a) Find $\text{cov}(X,Y)$. (b) Note that if we know X we know Y. Is $\rho(X,Y) = 1$?

5.10. Suppose U, V, W are independent and have mean μ and variance σ^2. Find $\rho(U + V, V + W)$.

5.11. Suppose that X_1, X_2, X_3 have mean 0 and $\text{var}(X_i) = i$. Find $\rho(X_1 - X_2, X_2 + X_3)$.

5.12. A purse contains three 10-cent and six 25-cent pieces. Pick two coins without replacement and let X_1 and X_2 be their values (measured in cents). (a) Find $\rho(X_1, X_2)$. (b) Would the answer change if we had three 1-cent and six 50-cent pieces instead?

5.13. Show that two random variables cannot possibly have the following properties: $EX = 1$, $EY = 3$, $EX^2 = 5$, $EY^2 = 10$, $EXY = 0$.

5.14. Suppose $EX = 1$, $EY = 3$, $EX^2 = 2$, $EY^2 = 13$, $EXY = 5$. What can we say about X and Y?

5.15. Suppose $\rho(X_1, X_2) = 1$, $EX_i = \mu_i$, and $\text{var}(X_i) = \sigma_i^2$. (5.3) tells us that $X_2 = aX_1 + b$. Find a and b.

5.16. Let X and Y be random variables with $EX = 2$, $\text{var}(X) = 1$, $EY = 3$, $\text{var}(Y) = 4$. What are the smallest and largest possible values of $\text{var}(X + Y)$?

5.17. Use the Cauchy-Schwarz inequality to show that for ANY random variables $\sigma(X + Y) \le \sigma(X) + \sigma(Y)$.

5.18. Suppose we draw n balls out of an urn that contains K red and L black balls and let U be the number of red balls and V be the number of black balls we obtain. Find $\rho(U, V)$.

5.19. Suppose $X_1, \ldots, X_n$ are such that $\text{var}(X_i) = \sigma^2$ and $\rho(X_i, X_j) = \rho$ when $i < j$. Then $\rho \ge -1/(n-1)$. Construct an example for which equality holds.

5.20. Suppose we draw k balls out of an urn that contains an equal number of balls of each of $n \ge 2$ colors. Let U_i be the number of balls of the ith color drawn. Find $\rho(U_i, U_j)$ for $i \ne j$.

5.21. m balls are drawn from an urn containing balls numbered $1, \ldots, n$. Find the mean and variance of the sum of the numbers drawn.

*4.6. Conditional Expectation

In this section we will consider the conditional mean and variance. These are nothing more than the mean and variance of the conditional distribution. In the discrete case the **conditional mean** is given by

(6.1) $$E(Y|X = x) = \sum_y y P(Y = y|X = x)$$

where $$P(Y = y|X = x) = \frac{P(X = x, Y = y)}{P(X = x)}$$

When X and Y have joint density f,

(6.2) $$E(Y|X = x) = \int y f_Y(y|X = x) \, dy$$

where $$f_Y(y|X = x) = \frac{f(x, y)}{f_X(x)}$$

As we have said repeatedly in Chapter 3, the second formula is obtained from the first by replacing the sum by an integral and the probabilities by the corresponding density functions. A concrete example should help explain (6.2).

Example 6.1. Suppose X and Y have joint density

$$f(x, y) = \begin{cases} 8xy & \text{if } 0 < y < x < 1 \\ 0 & \text{otherwise} \end{cases}$$

$$f_X(x) = \int f(x, y)\, dy = \int_0^x 8xy\, dy = 4xy^2 \Big|_0^x = 4x^3$$

$$f_Y(y|X = x) = \frac{f(x, y)}{f_X(x)} = \frac{8xy}{4x^3} = \frac{2y}{x^2} \qquad \text{for } 0 < y < x$$

$$E(Y|X = x) = \int y f_Y(y|X = x)\, dy = \int_0^x y\frac{2y}{x^2}\, dy = \frac{2y^3}{3x^2}\Big|_0^x = \frac{2x}{3}$$

Notice that $E(Y|X = x)$ is a function of x. For the moment, call that function $h(x)$. In the example, $h(x) = 2x/3$. We define $E(Y|X)$ to be the random variable $h(X)$. In the example $E(Y|X) = 2X/3$. In words, $E(Y|X = x)$ is a function that for each possible value x of X tells us the conditional mean of Y given that $X = x$, while $E(Y|X)$ is a random variable that tells us the conditional mean of Y for the value of X we have observed. The next formula and Example 6.2 will explain our interest in this concept.

(6.3) $E\{E(Y|X)\} = EY$

PROOF: Here and throughout this section we will give the proof for the the continuous case. The details in the discrete case are similar but simpler. Plugging in the definitions,

$$E\{E(Y|X)\} = \int E(Y|X = x) f_X(x)\, dx$$

$$= \int \int y\, \frac{f(x, y)}{f_X(x)}\, dy\, f_X(x)\, dx$$

$$= \int \int y f(x, y)\, dy\, dx = EY$$

by our formula for $Er(X, Y)$, (2.7).   □

Using (6.3) on Example 6.1 we have

$$EY = E\{E(Y|X)\} = \int E(Y|X = x) f_X(x)\, dx$$

$$= \int_0^1 \frac{2x}{3} 4x^3\, dx = \frac{8x^5}{15}\Big|_0^1 = \frac{8}{15}$$

To confirm this answer we note that

$$f_Y(y) = \int f(x,y)\, dx = \int_y^1 8xy\, dx = 4x^2 y\big|_y^1 = 4(y - y^3) \quad 0 < y < 1$$

$$EY = \int y f_Y(y)\, dy = \int_0^1 y\, 4(y - y^3)\, dy = \left(\frac{4y^3}{3} - \frac{4y^5}{5}\right)\bigg|_0^1 = \frac{8}{15}$$

Example 6.2. Suppose we want to determine the expected value of $Y =$ the number of rolls it takes to complete a game of craps. (See Example 3.4 in Chapter 2 for a description of the game.) Let X be the sum we obtain on the first roll. If $X = 2, 3, 7, 11, 12$ then the outcome is determined by the first roll so in these cases $E(Y|X = x) = 1$. If $X = 4$ then the game is completed when a 4 or a 7 appears. So we are waiting for an event with probability 9/36 and the formula for the mean of the geometric tells us that the expected number of rolls is 36/9. Adding the first roll, we have $E(Y|X = 4) = 45/9$. Similar calculations give us

x	2,3,7,11,12	4,10	5,9	6,8
$E(Y|X = x)$	1	45/9	46/10	47/11
probability	12/36	6/36	8/36	10/36

Putting the pieces together,

$$EY = E\{E(Y|X)\} = \sum_x E(Y|X = x)P(X = x)$$

$$= 1 \cdot \frac{12}{36} + \left(5 \cdot \frac{6}{36} + 4.6 \cdot \frac{8}{36} + \frac{47}{11} \cdot \frac{10}{36}\right)$$

$$= 0.333 + 0.833 + 1.022 + 1.187 = 3.375$$

Since the conditional expectation is just the expected value for the conditional distribution, it has the same properties that the ordinary expectation has. For example, to take the conditional expectation of a function we just integrate the function times the conditional distribution.

$$E(r(Y)|X = x) = \int r(y) f_Y(y|X = x)\, dy$$

(6.4) $$\qquad E(r(X, Y)|X = x) = \int r(x, y) f_Y(y|X = x)\, dy$$

$$E(r(Y, Z)|X = x) = \iint r(y, z) f_{Y,Z}(y, z|X = x)\, dy\, dz$$

As in Section 4.3, the last formula leads easily to the fact that the conditional expectation of the sum is the sum of the conditional expectations:

(6.5) $$E(Y + Z|X) = E(Y|X) + E(Z|X)$$

PROOF: $f_{Y,Z}(y, z|X = x) = f_{X,Y,Z}(x, y, z)/f_X(x)$, so using (6.4) and the fact that the integral of the sum is the sum of the integrals,

$$E(Y + Z|X = x) = \iint (y + z) f_{Y,Z}(y, z|X = x) \, dy \, dz$$
$$= \iint y \, f_{Y,Z}(y, z|X = x) \, dy \, dz + \iint z \, f_{Y,Z}(y, z|X = x) \, dy \, dz$$
$$= E(Y|X = x) + E(Z|X = x)$$

Since this result holds for each x, and $E(W|X)$ is $E(W|X = x)$ evaluated at $x = X$ the result follows.   □

The expected value satisfies $E(cY) = cEY$. For conditional expectation, more is true:

(6.6) $$E(h(X)Y|X) = h(X)E(Y|X)$$

In words, when we condition on the value of X, $h(X)$ is known and can be treated like a constant.

PROOF: As in the proof of (6.5) the result will be established once we show

$$E(h(X)Y|X = x) = h(x)E(Y|X = x)$$

By (6.1) of Chapter 3, the joint density of X and $Z = h(X)Y$ is (when $h(x) > 0$)

$$f_{X,Z}(x, z) = f_{X,Y}(x, z/h(x))/h(x)$$

so $f_Z(z|X = x) = f_{X,Y}(x, z/h(x))/\{h(x)f_X(x)\}$ and

$$E(h(X)Y|X = x) = \int z \, \frac{f_{X,Y}(x, z/h(x))}{h(x)f_X(x)} \, dz$$

Changing variables $y = h(x)z$, the last integral

$$= h(x) \int y \, \frac{f_{X,Y}(x, y)}{f_X(x)} \, dy = h(x)E(Y|X = x)$$

When $h(x) < 0$ the last calculation works with one small change. We have $|h(x)| = -h(x)$ in the denominator of $f_{Z|X}$ but the minus sign is canceled in the change of variables $y = h(x)z$ since we have to reverse the limits of integration. When $h(x) = 0$, $E(h(X)Y|X = x) = 0$ so the formula holds in this case as well. □

As an application of (6.6) we get

(6.7) If $E(Y|X)$ is constant, X and Y are uncorrelated.

PROOF: (6.3) implies $E\{E(Y|X)\} = EY$, so the constant must be EY. (6.6) implies

$$E(XY|X) = XE(Y|X) = XEY$$

So taking expected values and using (6.3), (6.6), and our assumption that $E(Y|X) = EY$,

$$E(XY) = E\{E(XY|X)\} = E\{XE(Y|X)\} = E\{XEY\} = EXEY \qquad □$$

EXERCISE 6.1. Give an example in which X and Y are uncorrelated but $E(Y|X)$ is not constant.

The **conditional variance** $\text{var}(Y|X = x)$ is the variance of the conditional distribution. Just as the ordinary variance satisfies

$$\text{var}(Z) = E(Z - EZ)^2 = EZ^2 - (EZ)^2$$

the conditional variance satisfies

$$\begin{aligned}
\text{var}(Y|X = x) &= E(\{Y - E(Y|X = x)\}^2|X = x) \\
&= E(Y^2|X = x) - E(Y|X = x)^2
\end{aligned}$$

the last term being the square of the conditional mean. To illustrate these formulas we compute the conditional variance for Example 6.1.

$$E(Y^2|X = x) = \int y^2 f_Y(y|X = x)\,dy = \int_0^x y^2 \frac{2y}{x^2}\,dy = \frac{2y^4}{4x^2}\Big|_0^x = \frac{x^2}{2}$$

$$E(Y|X = x)^2 = \left(\frac{2x}{3}\right)^2 = \frac{4x^2}{9}$$

$$\text{var}(Y|X = x) = E(Y^2|X = x) - E(Y|X = x)^2 = \frac{x^2}{2} - \frac{4x^2}{9} = \frac{x^2}{18}$$

Example 6.3. If X_1 and X_2 have the bivariate normal distribution described in Example 5.6 then X_1 and X_2 are related by

$$X_2 = \mu_2 + \rho \frac{\sigma_2}{\sigma_1}(X_1 - \mu_1) + \sigma_2 \sqrt{1 - \rho^2}\, Z$$

where Z is an independent standard normal. From this representation we can compute

$$E(X_2|X_1 = x_1) = \mu_2 + \rho \frac{\sigma_2}{\sigma_1}(x_1 - \mu_1)$$

$$\operatorname{var}(X_2|X_1 = x_1) = \sigma_2^2(1 - \rho^2)$$

EXERCISE 6.2. Suppose that the first midterm grades have a mean of 75 and variance of 25, the second midterm grades have a mean of 80.4 and variance of 16, and the two grades have a bivariate normal distribution with $\rho^2 = 0.75$. What is the probability that a student who got an 80 on the first midterm will get a higher grade on the second?

As in the case of the conditional mean, $\operatorname{var}(Y|X = x)$ is a function of x, which for the moment we can call $h(x)$, and we define a random variable $\operatorname{var}(Y|X)$ to be $h(X)$. Again, $\operatorname{var}(Y|X = x)$ is the function that for each possible value x of X tells us the conditional variance of Y given that $X = x$, while $\operatorname{var}(Y|X)$ is a random variable that tells us the conditional variance of Y for the value of X we have observed. Erasing the "$= x$" from the definition of $\operatorname{var}(Y|X = x)$, we have

$$(6.8) \qquad \begin{aligned} \operatorname{var}(Y|X) &= E(\{Y - E(Y|X)\}^2|X) \\ &= E(Y^2|X) - E(Y|X)^2 \end{aligned}$$

which leads us to the following important formula:

$$(6.9) \qquad \operatorname{var}(Y) = E\{\operatorname{var}(Y|X)\} + \operatorname{var}\{E(Y|X)\}$$

PROOF: (6.3) implies $E\{E(Y^2|X)\} = EY^2$ so taking the expected value of (6.8) we have

$$E\{\operatorname{var}(Y|X)\} = EY^2 - E\{E(Y|X)^2\}$$

$\operatorname{var}(Z) = EZ^2 - (EZ)^2$ so taking $Z = E(Y|X)$ and recalling that $E(Y|X)$ has mean EY by (6.3), we have

$$\operatorname{var}\{E(Y|X)\} = E\{E(Y|X)^2\} - (EY)^2$$

Adding the last two equations we have

$$E\{\operatorname{var}(Y|X)\} + \operatorname{var}\{E(Y|X)\} = EY^2 - (EY)^2 = \operatorname{var}(Y)$$

proving the result. □

Of course formula (6.9) holds in Example 6.1. Using the formulas $f_X(x) = 4x^3$ for $0 < x < 1$, $f_Y(y) = 4(y - y^3)$ for $0 < y < 1$, and $EY = 8/15$ found earlier,

$$EY^2 = \int_0^1 y^2\, 4(y - y^3)\, dy = \left(y^4 - \frac{4y^6}{6} \right)\Big|_0^1 = \frac{1}{3}$$

$$\text{var}(Y) = EY^2 - (EY)^2 = \frac{1}{3} - \frac{64}{225} = 11/225$$

$$EX = \int_0^1 x\, 4x^3\, dx = 4/5$$

$$EX^2 = \int_0^1 x^2\, 4x^3\, dx = 2/3$$

$$\text{var}(X) = 2/3 - 16/25 = 2/75$$

Recalling that $\text{var}(Y|X) = X^2/18$ and $E(Y|X) = 2X/3$, we have

$$E\{\text{var}(Y|X)\} = E(X^2/18) = 1/27 = 0.0370$$
$$\text{var}\{E(Y|X)\} = \text{var}(2X/3) = (4/9)\text{var}(X) = 8/(27 \cdot 25) = 0.0118$$

Adding the last two formulas, we see

$$E\{\text{var}(Y|X)\} + \text{var}\{E(Y|X)\} = \frac{1}{27} + \frac{8}{27 \cdot 25} = \frac{33}{27 \cdot 25} = \text{var}(Y)$$

Formula (6.9) is important in statistics. To explain this we recall that in Exercise 4.8 you showed

$$\text{var}(Y) = \min_c E(Y - c)^2$$

and the minimizing $c = EY$. Applying this reasoning to the conditional distribution shows that $\min_f E(Y - f(X))^2$ is attained when $f(x) = E(Y|X = x)$ so

(6.10) $$E\{\text{var}(Y|X)\} = \min_f E(Y - f(X))^2$$

Thus (6.9) splits the variance of Y into two pieces; the variance of our best guess at Y given X, $\text{var}\{E(Y|X)\}$, and the variance that remains after our best guess, $E\{\text{var}(Y|X)\}$.

Continuing to refer back to Example 6.1, we see in that case $E(Y|X) = 2X/3$, i.e., the optimal $f(X)$ is a linear function and hence agrees with the optimal linear predictor found in Example 5.5. Of course, the amounts of variance not explained by the line, which we calculated here as $E\{\mathrm{var}(Y|X)\}$ and there as $(1-\rho^2)\mathrm{var}(Y)$, are equal (to $1/27$). In general $E(Y|X)$ may be a nonlinear function of X so

$$E\{\mathrm{var}(Y|X)\} \leq (1-\rho^2)\mathrm{var}(Y)$$

(6.9) is useful in computing the variance in concrete examples.

Example 6.4. Random sums. Let $X_1, X_2, \ldots$ be independent random variables that all have the same distribution, let N be an independent integer-valued random variable, and let $S_N = X_1 + \cdots + X_N$. For a concrete example suppose that the number of people who visit a fast food restaurant during the lunch hour is a Poisson random variable N with mean 100 and suppose that each patron spends \$2 with probability 0.3, \$3 with probability 0.4, and \$4 with probability 0.3. Then S_N gives the total sales during the lunch hour. If $\mu = EX_i$ and $\sigma^2 = \mathrm{var}(X_i)$ then since N is independent of the X_i,

$$E(S_N|N=n) = E(X_1 + \cdots + X_n|N=n) = \mu n$$
$$\mathrm{var}(S_N|N=n) = \mathrm{var}(X_1 + \cdots + X_n|N=n) = \sigma^2 n$$

So $E(S_N|N) = \mu N$, $\mathrm{var}(S_N|N) = \sigma^2 N$, and using (6.3) and (6.9) we have

$$ES_N = E\{E(S_N|N)\} = \mu EN$$

(6.11)
$$\mathrm{var}(S_N) = E\{\mathrm{var}(S_N|N)\} + \mathrm{var}\{E(S_N|N)\}$$
$$= \sigma^2 EN + \mu^2 \mathrm{var}(N)$$

In the concrete example $\mu = 3$, $\sigma^2 = E(X-3)^2 = 0.6$, $EN = 100$, and $\mathrm{var}(N) = 100$, so

$$ES_N = 300 \quad \text{and} \quad \mathrm{var}(S_N) = (0.6)100 + (9)100 = 960$$

To confirm this answer we note that Example 5.10 in Chapter 3 implies that if N_k = the number of customers who spend \$$k$ then N_2, N_3, N_4 are independent Poissons with means 30, 40, and 30 and hence variances 30, 40, and 30. So recalling $\mathrm{var}(cN_k) = c^2\mathrm{var}(N_k)$, we have

$$\mathrm{var}(S_N) = 4 \cdot 30 + 9 \cdot 40 + 16 \cdot 30 = 960$$

A variance of 960 may sound large but taking the square root we find that the standard deviation is 30.98, a little more than 10% of the mean revenue, 300.

Example 6.5. One can also use (6.11) to compute the variance of the number of rolls to complete the game of craps, which we studied in Example 6.2. We begin by observing that if $X = \text{geometric}(p)$ then $X + 1$ has mean $1 + (1/p)$ and variance $(1 - p)/p^2$ so

x	2,3,7,11,12	4,10	5,9	6,8
p		9/36	10/36	11/36
$E(Y\|X = x)$	1	5	46/10	47/11
$\text{var}(Y\|X = x)$	0	$27(36)/9^2$	$26(36)/10^2$	$25(36)/11^2$
probability	12/36	6/36	8/36	10/36

Using the information in the table and the value of $E\{E(Y|X)\} = EY$ computed earlier,

$$E\{\text{var}(Y|X)\} = \frac{6 \cdot 27}{9^2} + \frac{8 \cdot 26}{10^2} + \frac{10 \cdot 25}{11^2} = 2 + 2.08 + 2.066 = 6.146$$

$$E\{E(Y|X)^2\} = \frac{1}{36}\left(1 \cdot 12 + 25 \cdot 6 + (4.6)^2 \cdot 8 + \left(\frac{47}{11}\right)^2 \cdot 10\right) = 14.273$$

$$\text{var}\{E(Y|X)\} = E\{E(Y|X)^2\} - (EY)^2 = 14.273 - (3.375)^2 = 2.882$$

$$\text{var}(Y) = E\{\text{var}(Y|X)\} + \text{var}\{E(Y|X)\} = 6.146 + 2.882 = 9.028$$

To check this answer we will compute it another way, which turns out to be a little simpler. The second moment of the geometric distribution is the variance plus the square of the mean:

$$(1 - p)/p^2 + (1/p)^2 = (2 - p)/p^2$$

So subtracting the first roll, we have

$$E\{E((Y - 1)^2|X)\} = \frac{6}{36} \cdot \frac{63 \cdot 36}{9^2} + \frac{8}{36} \cdot \frac{62 \cdot 36}{10^2} + \frac{10}{36} \cdot \frac{61 \cdot 36}{11^2}$$

$$= 4.666 + 4.960 + 5.041 = 14.668$$

and hence

$$\text{var}(Y) = \text{var}(Y - 1) = E(Y - 1)^2 - (E(Y - 1))^2$$
$$= 14.668 - (2.375)^2 = 9.027$$

which is the same answer except for a little round-off error.

Example 6.6. Generating functions of random sums. Suppose $X_1, X_2, X_3, \ldots$ are i.i.d. and N is an independent integer-valued random variable. For example, if we catch a random number of fish N, and the ith weighs X_i pounds, then $S_N = X_1 + \cdots + X_N$ is the total weight of the fish caught. To compute the m.g.f. of S_N we note that

$$E(\exp(tS_N)|N = n) = E(\exp(t(X_1 + \cdots + X_n)|N = n)$$
$$= \phi_X(t)^n = e^{n \ln(\phi_X(t))}$$

Taking expected values and using (6.3),

$$E \exp(tS_N) = \sum_n E(\exp(tS_N)|N = n)P(N = n)$$

(6.12)
$$= \sum_n e^{n \ln(\phi_X(t))} P(N = n)$$

$$= E\{\exp(N \ln \phi_X(t))\} = \phi_N(\ln \phi_X(t))$$

EXERCISE 6.3. The calculation in (6.12) comes out a little neater if we look at the generating function. Suppose that the X_i are nonnegative integer-valued and show that $Ez^{S_N} = \gamma_N(\gamma_X(z))$.

EXERCISE 6.4. Differentiate (6.12) twice to show that

$$ES_N = EN\,EX \qquad E(S_N)^2 = EN^2 EX + EN\,\text{var}(X)$$

and get a new proof of (6.11).

Example 6.7. An important special case occurs when N is Poisson(λ). In this case, S_N is said to have a **compound Poisson distribution**. Combining (6.12) with the formula from Example 2.4, for the moment generating function of the Poisson(λ), $\phi_N(t) = \exp(\lambda(e^t - 1))$, we have

$$E(\exp(tS_N)) = \exp(\lambda(\phi_X(t) - 1))$$

If we suppose further that the X_i have a Bernoulli distribution then $\phi_X(t) = (1-p)+pe^t$ so the sum of a Poisson(λ) number of Bernoulli(p) random variables has

$$E(\exp(tS_N)) = \exp(\lambda p(e^t - 1))$$

that is, S_N is Poisson(λp).

EXERCISES

6.5. Suppose we roll a die to get a number and then flip a coin that number of times. What is the expected number of Heads that we will get?

6.6. Suppose that a point X is chosen uniformly on the interval $(0,1)$ and then Y is chosen uniformly on $(0, X)$. Find EY.

6.7. Suppose we roll a die n times and let X be the number of odd numbers that appear and Y be the number of 6's. Find $E(X|Y)$ and $E(Y|X)$.

6.8. Suppose $E(X|Y) = 18 - 3Y/5$ and $E(Y|X) = 10 - X/3$. (This is the answer to the last problem with $n = 30$.) Use (6.3) to find EX and EY.

6.9. Suppose $X_1, \ldots, X_n$ are independent and have the same distribution. Find $E(X_1 + \cdots + X_m | X_1 + \cdots + X_n)$.

6.10. What is the expected number of Heads in the first four tosses given that we got 7 Heads in the first 10 tosses?

6.11. Suppose $X = \text{binomial}(n, p)$ and $Y = \text{binomial}(m, p)$ are independent, and let $Z = X + Y$. Find $E(X|Z)$.

6.12. Suppose that the number of customers who arrive at a fast food restaurant in nonoverlapping time intervals of length one minute are independent Poisson random variables with mean 5/3. What is the conditional expectation of the number of people who arrived between 12:00 and 12:15 given that 112 people arrived between 12:00 and 1:00?

6.13. Suppose $X = \text{Poisson}(\lambda)$ and $Y = \text{Poisson}(\mu)$ are independent, and let $Z = X + Y$. Find $E(X|Z)$ and $\text{var}(X|Z)$.

6.14. Suppose X and Y have joint density $(2m + 6)x^m y$ for $0 < y < x < 1$. Find $E(Y|X)$, and $\text{var}(Y|X)$. Note that the answers do not depend on m.

6.15. Suppose X and Y have joint density e^{-x} for $0 < y < x$. Find $E(X|Y)$ and $\text{var}(X|Y)$.

6.16. Let U_1, U_2, U_3 be independent and uniform on $(0,1)$. If we let $X = \min\{U_1, U_2, U_3\}$ and $Y = \max\{U_1, U_2, U_3\}$ then the joint density of X and Y is $f(x, y) = 6(y - x)$ if $0 < x < y < 1$ and 0 otherwise. Compute $E(X|Y)$ and $\text{var}(X|Y)$.

6.17. Suppose we flip a coin repeatedly until we get Tails on two successive tosses. Let T be the number of tosses required. (a) By considering the outcomes of the first and second tosses, argue that

$$ET = \frac{1}{2}(1 + ET) + \frac{1}{4}(2 + ET) + \frac{1}{4}(2)$$

and solve to show that $ET = 6$. (b) Is this the same as the expected waiting time for HT?

6.18. Find the distribution of the duration of the game of craps.

6.19. Branching processes. Suppose $N_0 = 1$ and define N_n inductively as follows

$$N_n = \begin{cases} \sum_{m=1}^{N_{n-1}} \xi_{n,m} & \text{if } N_{n-1} > 0 \\ 0 & \text{if } N_{n-1} = 0 \end{cases}$$

where the $\xi_{n,m}$ are independent and have $P(\xi_{n,m} = k)$. Here we are thinking of N_n as the number of males in the nth generation and we suppose that each man has an independent and identically distributed number of male children. (a) Let $\mu = E\xi_{n,m}$. Use the observation

$$EN_n = E\{E(N_n|N_{n-1})\} = E\{\mu N_{n-1}\}$$

and induction to show that $EN_n = \mu^n$ where $\mu = EN_1$. From this it follows that if $\mu < 1$, $P(N_n > 0) \le EN_n$ goes to 0 exponentially fast, while if $\mu > 1$, EN_n grows exponentially rapidly. (b) Let $\sigma^2 = \text{var}(\xi_{n,m})$. Use (6.11) and (b) to conclude that

$$\text{var}(N_n) = \sigma^2 \mu^{n-1} + \mu^2 \text{var}(N_{n-1})$$

and hence $\text{var}(N_n) = \sigma^2 \sum_{k=n-1}^{2n-2} \mu^m$.

6.20. Consider the branching process defined in the previous problem. The goal of this problem, which was the motivation of Galton and Watson, two 19th century clergymen who introduced the process, is to compute $\lim_{n \to \infty} P(Z_n = 0)$ = the probability that the family line dies out. Let $\phi(z) = \sum_{k=0}^{\infty} p_k z^k$ and ϕ_n be the generating function of N_n. Since $N_0 = 1$, $\phi_1 = \phi$. (a) Use the observation that

$$E(z^{N_n}) = E\{E(z^{N_n}|N_1)\}$$

to show that $\phi_n(z) = \phi(\phi_{n-1}(z))$. (b) Show that as $n \to \infty$, $\phi_n(0) = P(Z_n = 0)$ increases to a limit x_0, which must be the smallest solution of $\phi(x) = x$.

6.21. Lotka observed that the number of male children in an American family was well approximated by the distribution with $p_k = bc^k$ for $k \ge 1$ and $p_0 = a = 1 - \{b/(1-c)\}$, where $b = 0.2126$ and $c = 0.5893$. Use the results of the last problem to calculate the probability that the branching process dies out for this offspring distribution.

4.7. Chapter Summary and Review Problems

Section 4.1. The expected value of X is defined to be

(1.1) $$EX = \sum_x x P(X = x)$$

(1.2) $$EX = \int x f(x)\,dx$$

assuming that $\sum_x |x| P(X = x) < \infty$ (or $\int |x| f(x)\, dx < \infty$). This quantity has a meaning much like the frequency intrepretation of probability: if $X_1, \ldots, X_n$ are independent and n is large then $(X_1 + \cdots + X_n)/n$ will be close to EX. (This is the weak law of large numbers that we will prove in Section 5.1.)

Section 4.2. To compute the expected value of a function of one (or more) random variable, we use the following formulas

(2.1) $$Er(X) = \sum_x r(x)P(X = x)$$

(2.2) $$Er(X) = \int r(x) f_X(x)\, dx$$

(2.6) $$Er(X, Y) = \sum_{x,y} r(x, y)P(X = x, Y = y)$$

(2.7) $$Er(X, Y) = \iint r(x, y) f_{X,Y}(x, y)\, dy\, dx$$

EX^k is called the **kth moment** of X. $\phi_X(t) = Ee^{tX}$ is called the **moment generating function** (or m.g.f.) and can be used to compute the moments of X.

(2.3) Suppose $\phi_X(t) = Ee^{tX} < \infty$ for $t \in (-t_0, t_0)$ for some $t_0 > 0$, and let $\phi^{(n)}$ denote the nth derivative of ϕ. Then

$$\phi^{(n)}(0) = EX^n$$

In **Section 4.3** we learned that

(3.2) $$\phi_{aX+b}(t) = e^{tb}\phi_X(ta)$$

and that if $X_1, \ldots, X_n$ are independent then

(3.9) $$\phi_{X_1 + \cdots + X_n}(t) = \phi_{X_1}(t) \cdots \phi_{X_n}(t)$$

This result allows us to compute the distribution of $X_1 + \cdots + X_n$ because of

(3.10) Suppose two m.g.f.'s $\phi_X(t)$ and $\phi_Y(t)$ are finite and equal for $t \in (-t_0, t_0)$ with $t_0 > 0$. Then X and Y have the same distribution.

Some important properties of expected value are:

(3.1) $E(aX + b) = aEX + b$

(3.4) If $X \geq Y$ then $EX \geq EY$.

(3.5) $E(X_1 + \cdots + X_n) = EX_1 + \cdots + EX_n$

If $X_1, \ldots, X_n$ are independent then

(3.7) $$E(X_1 \cdots X_n) = EX_1 \cdots EX_n$$

but the equality may hold for random variables that are not independent. Random variables X and Y that have $EXY = EXEY$ are said to be **uncorrelated**.

Section 4.4. The **variance** of X,

$$\text{var}(X) = E\{(X - EX)^2\} = EX^2 - (EX)^2$$

The first equality is the definition and the second is (4.1). The variance satisfies

(4.2)· $$\text{var}(aX + b) = a^2\text{var}(X)$$

a relationship that shows us that the variance is a measure of how spread-out the distribution is. The a^2 in the last relationship motivates us to define the **standard deviation** $\sigma(X) = \sqrt{\text{var}(X)}$, which satisfies

(4.4) $$\sigma(aX + b) = |a|\sigma(X)$$

The **covariance** of X,

$$\text{cov}(X) = E\{(X - EX)(Y - EY)\} = EXY - EX\,EY$$

The first equality is the definition and the second is (4.5). In general

(4.6) $$\text{var}\left(\sum_{i=1}^n X_i\right) = \sum_{i=1}^n \text{var}(X_i) + 2 \sum_{1 \leq i < j \leq n} \text{cov}(X_i, X_j)$$

If the X_i are independent, or more generally, have $\text{cov}(X_i, X_j) = 0$ when $i \neq j$, then the last sum disappears and the variance of the sum is the sum of the variances.

Section 4.5. If X and Y are random variables with standard deviations $\sigma(X)$ and $\sigma(Y)$ that are positive we define the **correlation** to be

$$\rho(X, Y) = \frac{\text{cov}(X, Y)}{\sigma(X)\sigma(Y)} = E\left\{\frac{(X - EX)}{\sigma(X)} \cdot \frac{(Y - EY)}{\sigma(Y)}\right\}$$

In words, we subtract the mean and divide by the standard deviation of each variable to make their means 0 and variances 1, then we take the expected value of the product. Since we subtract the mean and divide by the standard deviation before taking the expectation, the correlation is not affected by a linear transformation in which a_1 and a_2 are positive.

(5.1) If $Y_1 = a_1 X_1 + b_1$ and $Y_2 = a_2 X_2 + b_2$ where $a_1, a_2 \neq 0$ then

$$\rho(Y_1, Y_2) = \frac{a_1}{|a_1|} \frac{a_2}{|a_2|} \rho(Y_1, Y_2)$$

In particular, $\rho(X, -Y) = -\rho(X, Y)$.

The **Cauchy-Schwarz inequality**, $(EUV)^2 \leq EU^2 EV^2$, implies that the correlation satisfies

(5.3) $-1 \leq \rho(X, Y) \leq 1$ with $|\rho(X, Y)| = 1$ if and only if $Y = aX + b$

For a quantitative interpretation of $\rho(X_1, X_2)$ we observe that the values of a and b that minimize $E(X_2 - (aX_1 + b))^2$ are

$$a = \rho \frac{\sigma_2}{\sigma_1} \qquad b = \mu_2 - a\mu_1$$

where $\mu_i = EX_i$ and $\sigma_i^2 = \text{var}(X_i)$. For these values the fraction of the variance of X_2 not explained by the line $aX_1 + b$ is

$$\frac{E(X_2 - (aX_1 + b))^2}{\sigma_2^2} = 1 - \rho^2$$

Section 4.6. The conditional mean of Y given $X = x$ is just the mean of the conditional distribution:

(6.1) $$E(Y|X = x) = \sum_y y \, P(Y = y|X = x)$$

(6.2) $$E(Y|X = x) = \int y \, f_Y(y|X = x) \, dy$$

where

$$P(Y = y|X = x) = \frac{P(X = x, Y = y)}{P(X = x)} \qquad f_Y(y|X = x) = \frac{f(x, y)}{f_X(x)}$$

If $h(x) = E(Y|X = x)$ then $h(X)$ is a random variable we call $E(Y|X)$. It tells us the conditional mean of Y for the value of X we have observed. This quantity has the following properties:

(6.3) $$E\{E(Y|X)\} = EY$$

(6.5) $$E(Y + Z|X) = E(Y|X) + E(Z|X)$$

(6.6) $$E(h(X)Y|X) = h(X)E(Y|X)$$

The conditional variance $\text{var}(Y|X = x)$ is the variance of the conditional distribution. The associated random variable $\text{var}(Y|X)$ can be defined by

(6.8) $$\text{var}(Y|X) = E(Y^2|X) - \{E(Y|X)\}^2$$

and satisfies

(6.9) $$\text{var}(Y) = E\{\text{var}(Y|X)\} + \text{var}\{E(Y|X)\}$$

The first term on the right,

(6.10) $$E\{\text{var}(Y|X)\} = \min_f E(Y - f(X))^2$$

with the minimum achieved by $f(X) = E(Y|X)$. So (6.9) splits the variance of Y into two pieces; the variance of our best guess at Y given X, $\text{var}\{E(Y|X)\}$, and the variance that remains after we guess, $E\{\text{var}(Y|X)\}$.

EXERCISES

7.1. A lottery has one \$100 prize, two \$25 prizes, and five \$10 prizes. What should you be willing to pay for a ticket if 100 tickets are sold?

7.2. A parking garage charges \$2 for the first hour and \$1 per hour (or fraction thereof) after that. Suppose we park in this lot for an exponentially distributed amount of time with mean 1. What is the expected cost to park?

7.3. An unreliable clothes dryer dries your clothes and takes 10 minutes with probability 0.6, buzzes for 2 minutes and does nothing with probability 0.3, and buzzes for three minutes and does nothing with probability 0.1. If we assume that successive trials are independent and that we patiently keep putting our money in to try to get it to work, what is the expected time we need to get our clothes dry?

7.4. Suppose we flip a coin repeatedly. Let W_n be the number of tosses we need to get n Heads in a row, and let V_n be the number of tosses we need to get $n-1$ Heads followed by a Tail. Compute (a) EW_n, (b) EV_n. The reason for interest in these particular waiting times is that if you write down your favorite string of Heads and Tails of length n and let T_n be the number of tosses until you see that string then $EV_n \le ET_n \le EW_n$.

7.5. Suppose X has distribution function $F(x) = 1 - (1+x)^{-2}$ for $x \ge 0$. Find EX and EX^2.

7.6. If X and Y are independent standard normals then $R = \sqrt{X^2 + Y^2}$ has the **Rayleigh distribution**. That is, R has density function $f(r) = re^{-r^2/2}$ for $x \ge 0$, 0 otherwise. Find the mean and variance of R.

7.7. Suppose X and Y are independent and uniform on $(0,1)$. Then $Z = XY$ has density function $f(z) = \ln(1/z)$ for $0 < z < 1$ and 0 otherwise. Find EZ by computing $\int z f(z)\, dz$ and then use $EZ = EXEY$ to check your answer.

7.8. Suppose $X_1, \ldots, X_n$ are independent and uniform on $(0,1)$. Let $Y = \min\{X_1, \ldots, X_n\}$ and $Z = \max\{X_1, \ldots, X_n\}$. The joint density of Y and Z is given by

$$f(y, z) = \begin{cases} n(n-1)(z-y)^{n-2} & \text{when } 0 < y < z < 1 \\ 0 & \text{otherwise} \end{cases}$$

Find the mean, second moment, and variance of $Z - Y$.

7.9. Suppose X and Y are independent with $\text{var}(X) = 5$ and $\text{var}(Y) = 2$. Find $\text{var}(2X - 3Y + 11)$.

7.10. Suppose X and Y are independent with $\text{var}(X) = a$ and $\text{var}(Y) = b$. Find θ to minimize the variance of $Z = \theta X + (1 - \theta)Y$. Note that we do not suppose EX and EY are 0, but we can without loss of generality.

7.11. Suppose X takes values in $(-1,1)$. What is the largest possible value of $\text{var}(X)$ and when is this attained? What is the answer when $(-1, 1)$ is replaced by (a, b)?

7.12. Suppose X and Y are independent random variables with positive finite variance. Show that $\text{var}(XY) = \text{var}(X)\text{var}(Y)$ if and only if $EX = EY = 0$.

7.13. Suppose $X_1, \ldots, X_n$ are independent with $EX_i = \mu$ and $\text{var}(X_i) = \sigma^2$. Find $E(X_1 + \cdots + X_n)^2$.

7.14. Let X have mean μ and variance σ^2. The **skewness** β_1 of a distribution is $\sqrt{\beta_1} = E(X - \mu)^3/\sigma^3$. Compute the skewness for (a) $X = $ exponential(λ), (b) $X = $ normal(μ, σ^2).

7.15. Let X have mean μ and variance σ^2. The **kurtosis** β_2 is defined by $\beta_2 = E(X - \mu)^4/\sigma^4$. Compute the kurtosis when (a) $P(X = 1) = P(X = -1) = 1/2$, (b) $X = $ normal(μ, σ^2).

7.16. Suppose we toss a fair coin n times. A run is a sequence of tosses that all have the same result. For example the sequence HHTHHHTTH has five runs. Compute the mean and variance of the number of runs.

7.17. Suppose we toss a fair coin n times. Let X be the number of Heads and Y the number of runs. Compute (a) $E(X|Y)$, (b) $\text{cov}(X,Y)$. (c) Are X and Y independent?

7.18. The game of pool uses fifteen balls numbered 1 to 15 (and an all-white ball that we will ignore in this problem). Suppose we arrange the fifteen balls in a line in a random order. What are the mean and variance of the number of balls between the balls numbered 1 and 15?

7.19. A cereal company advertises that in each box is one playing card with a picture of a member of the 1992 U.S. Olympic basketball "dream team." What are the mean and variance of the number of boxes will you will have to buy to get all 10 cards?

7.20. Four men and four women are randomly assigned to four two-person canoes. Find the mean and variance of the number of canoes with one man and one woman.

7.21. Suppose U, V, and W are independent and uniform on (0,1). Find $\text{var}(UV)$, $\text{cov}(UV,(1-V)W)$.

7.22. Define a quadrilateral Q by picking a point uniformly distributed on each side of the unit cube $[0,1] \times [0,1]$. (More formally, we let U_1, U_2, U_3, U_4 be independent uniform on (0,1) and define Q to have corners at $(U_1, 0)$, $(1, U_2)$, $(1-U_3, 1)$, and $(0, 1-U_4)$.) Find the mean and variance of A = the area of Q.

7.23. Suppose U is uniform on (a) $[0, \pi/2]$, (b) $[0, \pi]$. Compute the covariance of $X = \sin(U)$ and $Y = \cos(U)$. Note that in either case $Y = \sqrt{(1 - X^2)}$.

7.24. Suppose X has mean 3 and variance 1, while Y has mean -2 and variance 4. Find the covariance of $X + Y$ and $X - Y$.

7.25. Suppose X and Y have joint density $f(x,y) = (y^2/12)e^{-x}$ if $x > 0$ and $-y < x < y$. Are X and Y (a) independent? (b) uncorrelated?

7.26. Suppose we roll a die repeatedly and let X be the number of the roll on which a 1 first appears and Y the number of the roll on which a 6 first appears. Show that conditional on $\{X < Y\}$, X and $Y - X$ are independent and have geometric distributions with parameters 1/3 and 1/6. Compute $\text{cov}(X,Y)$.

7.27. A hat contains ten tickets numbered 1, 5, 8, 15, 18, 58, 158, 581, 644, and 815. Let X and Y be the numbers on two tickets drawn without replacement. Find $\rho(X,Y)$.

5 Limit Theorems

5.1. Laws of Large Numbers

If $X_1, X_2, \ldots$ are independent and have the same distribution then we say the X_i are **independent and identically distributed**, or i.i.d. for short. Such sequences arise if we repeat some experiment such as flipping a coin or rolling a die, or if we stop people at random and measure their height or ask them how they will vote in an upcoming election. As we indicated at the beginning of Chapter 4, if $X_1, X_2, \ldots$ are i.i.d. with $EX_i = \mu$ then when n is large, the average of the first n observations, $(X_1 + \cdots + X_n)/n$, will be close to EX with high probability. Our first goal in this section will be to prove this result, which is called the **weak law of large numbers**. To prove the "weak law," we will use an inequality that says "if the variance is small then the random variable is close to its mean." That inequality, (1.2), is a special case of a simple general result.

(1.1) Markov's inequality. Suppose ϕ is nonnegative and nondecreasing on $[0, \infty)$ and $X \geq 0$. Then when $\phi(y) > 0$,

$$P(X \geq y) \leq E\phi(X)/\phi(y)$$

PROOF: Let $1_{\{X \geq y\}}$ be the random variable that is 1 if $X \geq y$ and 0 otherwise. This notation allows us to write

$$\phi(X) \geq \phi(y)1_{\{X \geq y\}}$$

To check this we note that when $X \geq y$, $\phi(X) \geq \phi(y)$ since ϕ is nondecreasing, while if $X < y$, $\phi(X) \geq 0$. Taking expected value and using the fact that if $Y \geq Z$ then $EY \geq EZ$ ((3.4) in Chapter 4), we have

$$E\phi(X) \geq E(\phi(y)1_{\{X \geq y\}}) = \phi(y)P(X \geq y)$$

Dividing both sides by $\phi(y)$, which we have assumed to be positive, gives the desired result. □

(1.2) Chebyshev's inequality. If $y > 0$ then

$$P(|Y - EY| \geq y) \leq \text{var}(Y)/y^2$$

PROOF: Taking $X = |Y - EY|$ and $\phi(x) = x^2$ in (1.1) gives

$$P(|Y - EY| \geq y) \leq E|Y - EY|^2/y^2 = \text{var}(Y)/y^2 \qquad □$$

Before applying (1.2) we will introduce some terminology. Let

$$\bar{X}_n = (X_1 + \cdots + X_n)/n$$

$\bar{X}_n$ is called the **sample mean** because if we assigned probability $1/n$ to each of the first n observations then $\bar{X}_n$ would be the mean of that distribution. If we suppose that the X_i are i.i.d. with $EX_i = \mu$ then using the facts that $E(cY) = cEY$ and the expected value of the sum is the sum of the expected values, we have

(1.3)
$$EX_n = \frac{1}{n}E(X_1 + \cdots + X_n)$$
$$= \frac{1}{n}\{EX_1 + \cdots + EX_n\} = \mu$$

If we suppose that $\text{var}(X_i) = \sigma^2$ then using the facts that $\text{var}(cY) = c^2\text{var}(Y)$ and that for independent $X_1, \ldots, X_n$ the variance of the sum is the sum of the variances, we have

(1.4)
$$\text{var}(\bar{X}_n) = \frac{1}{n^2}\text{var}(X_1 + \cdots + X_n)$$
$$= \frac{1}{n^2}\{\text{var}(X_1) + \cdots + \text{var}(X_n)\} = \frac{\sigma^2}{n}$$

Combining the last observation with Chebyshev's inequality, we see that if $\epsilon > 0$ then

(1.5)
$$P(|\bar{X}_n - \mu| \geq \epsilon) \leq \frac{\text{var}(\bar{X}_n)}{\epsilon^2} = \frac{\sigma^2}{\epsilon^2 n} \to 0$$

as $n \to \infty$. It is convenient to give a name to the conclusion we have just proved. We say that a sequence of random variables Z_n **converges to** b **in**

probability if for any $\epsilon > 0$ we have $P(|Z_n - b| > \epsilon) \to 0$ as $n \to \infty$. Using our new terminology, (1.5) can be expressed succinctly as "$\bar{X}_n$ converges to μ in probability." This brings us to

(1.6) Weak law of large numbers. Suppose $X_1, X_2, \ldots$ are i.i.d. with $E|X_i| < \infty$. Then as $n \to \infty$, $\bar{X}_n$ converges to EX_i in probability.

Careful readers will have noticed that we have stated the result assuming only that the mean exists, but proved it assuming that the variance is finite. We will not indicate how one treats the situation in which $E|X_i| < \infty$ but $\text{var}(X_i) = \infty$ since the details of the proof get much more complicated and the gain of generality is not very significant – in almost all applications $\text{var}(X_i) < \infty$.

Example 1.1. (1.6) could be easily called the fundamental theorem of statistics because it says that the sample mean is close to the mean μ of the underlying population when the sample is large. This implies, for example, that if we want to determine the average height of the 21-year-old American male, we do not have to measure the heights of all of the more than one million people in this category, but can instead estimate this quantity by measuring the heights of say 1,000 such individuals chosen at random. Later in this section and in the next one, we will deal with the important question: How close will the mean of a sample of size 1,000 be to the mean of the underlying population μ?

Example 1.2. As an application of (1.6) we get a justification of the frequency interpretation of probability. Suppose we repeat an experiment a large number of times and let $X_i = 1$ if A occurs on the ith trial and 0 otherwise. In this case $X_1 + \cdots + X_n$ is the number of times A occurs in the first n trials and $\bar{X}_n$ is the fraction of times A occurred. (1.6) implies that if n is large the frequency $\bar{X}_n$ is close to $EX_i = P(A)$ with high probability.

The conclusion of (1.6) is called "weak" because it does not rule out the possibility that the sequence of sample means $\bar{X}_1, \bar{X}_2, \ldots$ stays close to EX most of the time but occasionally wanders off because of a streak of bad luck. Our next result says that this does not happen.

(1.7) Strong law of large numbers. Suppose $X_1, X_2, \ldots$ are i.i.d. with $E|X_i| < \infty$. Then with probability one the sequence of numbers $\bar{X}_n$ converges to EX_i as $n \to \infty$.

Figure 5.1 illustrates (1.7) by showing the difference between the fraction of Heads in the first n tosses and $1/2$ for $10 \le n \le 5000$. Note that the rate of convergence to 0 is rather slow.

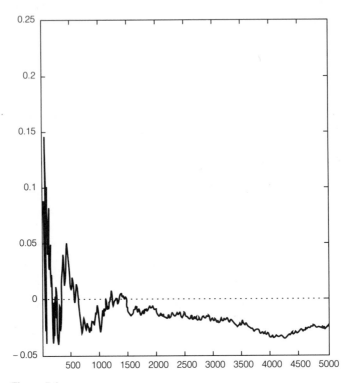

Figure 5.1

The first thing we have to explain is the phrase "with probability one." To do this we first consider flipping a coin and letting X_i be 1 if the ith toss results in Heads, and 0 otherwise. (1.7) says that with probability one

$$(X_1 + \cdots + X_n)/n \to 1/2 \qquad \text{as } n \to \infty$$

but in this case it is easy to write down sequences of tosses for which this is false:

$$
\begin{array}{ccccccccccc}
H & H & H & H & H & H & H & H & H & \cdots \\
H & H & T & H & H & T & H & H & T & \cdots
\end{array}
$$

The strong law of large numbers implies that the collection of "bad sequences" (i.e., those for which the asymptotic frequency of Heads is not 1/2) has probability zero. We will not prove the strong law of large numbers here since (i) the proof is rather difficult (even if we suppose $EX_i^2 < \infty$) and (ii) we will not live long enough to see the entire infinite sequence of X_i's, so the weak law, which says that for large n we are close to μ with high probability, is enough for most applications.

A nice feature of the proof of the weak law of large numbers is that it gives us a reasonable estimate, (1.5), of how close $\bar{X}_n$ is to μ. The next example illustrates this point.

Example 1.3. Suppose we flip a coin $n = 10,000$ times and let X_i be 1 if the ith toss is Heads and 0 otherwise, so that $\bar{X}_n$ is the fraction of the first n tosses that result in Heads. The X_i are Bernoulli random variables with $p = P(X_i = 1) = 1/2$, so $EX_i = p = 1/2$ and $\text{var}(X_i) = p(1-p) = 1/4$. Taking $\epsilon = 0.01$, (1.5) implies that

$$P(|\bar{X}_n - 1/2| \geq 0.01) \leq \frac{1/4}{(0.01)^2 10^4} = \frac{1}{4}$$

It is easier to see what is happening in the last computation if it is done in greater generality. Setting $\epsilon = c\sigma n^{-1/2}$ in (1.5) gives

(1.8) $$P(|\bar{X}_n - \mu| \geq c\sigma n^{-1/2}) \leq \frac{\sigma^2}{(c\sigma n^{-1/2})^2 n} = \frac{1}{c^2}$$

In words, the probability $\bar{X}_n$ differs from its mean μ by more than a constant c times $\sigma n^{-1/2}$, the standard deviation of $\bar{X}_n$, is smaller than $1/c^2$. In the next section we will see that when $\sigma^2 < \infty$ the difference between $\bar{X}_n$ and μ is typically a small multiple of $\sigma n^{-1/2}$ but the bound in (1.8) considerably overestimates the probability of error. In Chebyshev's defense we would like to observe that his bound is simple to prove and valid for all n and c, while the results in the next section are harder to prove and only hold for large n.

EXERCISES

1.1. Let X be exponential with mean 1. (a) Use Chebyshev's inequality to estimate $P(X > k)$ and compare this with the exact answer. (b) Take $k = 4$ in part (a).

1.2. Let $X = \text{binomial}(4, 1/2)$. Use Chebyshev's inequality to estimate $P(|X - 2| \geq 2)$ and compare with the exact probability.

1.3. Let Y be a standard normal and compare the upper bound on $P(|Y| \geq 3)$ that comes from Chebyshev's inequality with the exact probability from the table.

1.4. Let Y be a standard normal and compare the upper bound on $P(|Y| \geq 6)$ that comes from Chebyshev's inequality with the upper bound proved in Exercise 2.14 in Chapter 3, $P(|Y| \geq y) \leq (e^{-y^2/2})/y$.

1.5. Let $y \geq \sigma$. Give an example of a random variable Y with $EY = 0$, var$(Y) = \sigma^2$, and $P(Y \geq y) = \sigma^2/y^2$, the upper bound in Chebyshev's inequality.

1.6. The last problem shows that Chebyshev's inequality cannot be improved. The point of this problem is to demonstrate that it is always bad. Exercise 2.27 in Chapter 4 applied to $|X - EX|$ implies that

$$E(X - EX)^2 = \int_0^\infty 2x P(|X - EX| > x)\, dx$$

Use this to conclude that if var$(X) < \infty$ then

$$\frac{P(|X - EX| > y)}{\text{var}(X)/y^2} \to 0 \qquad \text{as } y \to \infty$$

In words, the true probability that $|X - EX| > y$ is always much smaller than the Chebyshev upper bound var$(X)/y^2$ when y is large.

1.7. Use (1.1) to prove **Bernstein's inequality**. For any $t > 0$

$$P(X \geq x) \leq e^{-tx} E e^{tX}$$

1.8. Apply Bernstein's inequality to $X = \text{Poisson}(\lambda)$ with $t = \ln c$ and $c > 1$ to conclude that

$$P(X > c\lambda) \leq \exp(\lambda\{c - 1 - c \ln c\})$$

The right-hand side is complicated but the derivative of $x - 1 - x \ln x$ is $-\ln x < 0$ for $x > 1$ so our inequality shows that if $c > 1$ then $P(X > c\lambda) \to 0$ exponentially fast as $\lambda \to \infty$.

1.9. Suppose $X_1, \ldots, X_n$ are independent exponential(1) random variables, and let $S_n = X_1 + \cdots + X_n$. Apply Bernstein's inequality to estimate $P(S_n > cn)$ with $c > 1$, then pick t to minimize the upper bound to show

$$P(S_n > cn) \leq e^{-n(c - 1 - \ln c)}$$

Again $P(S_n > cn) \to 0$ exponentially fast as $n \to \infty$.

1.10. Suppose $X_1, X_2, \ldots$ are independent and uniform on $(0,1)$. Let $m_n = \min\{X_1, \ldots, X_n\}$. (a) Show that $m_n \to 0$ in probability. (b) Show that $P(m_n > x/n) \to e^{-x}$.

1.11. Suppose $X_1, X_2, \ldots$ are independent and have an exponential distribution with parameter 1. Let $M_n = \max\{X_1, \ldots, X_n\}$. Show that if $a < 1 < b$ then $P(M_n \leq a \ln n) \to 0$ and $P(M_n \leq b \ln n) \to 1$ so $M_n / \ln n \to 1$ in probability as $n \to \infty$.

1.12. Normal numbers. Pick a point U uniform on (0,1) and let $X_1, X_2, \ldots$ be the successive digits in its decimal expansion. (a) Show that the X_i are independent and uniform on $\{0, 1, \ldots, 9\}$. (b) Use the strong law of large numbers to conclude that with probability one the fraction of time each digit appears in the first n places of the decimal expansion of U converges to $1/10$ as $n \to \infty$. Emile Borel called such numbers normal. We have shown that with probability one a number chosen at random from (0,1) will be normal. Unfortunately, this does not tell us anything about a specific number such as $\pi - 3$ or $e - 2$ that we might be interested in.

1.13. Monte Carlo integration. Suppose r has $\int_0^1 |r(x)| \, dx < \infty$ and let $U_1, U_2, \ldots$ be i.i.d. uniform on (0,1). Show that with probability one $I_n = (r(U_1) + \cdots + r(U_n))/n$ converges in probability to $I = \int_0^1 f(x) \, dx$.

1.14. Renewal theory. Suppose $t_1, t_2, \ldots$ are i.i.d. with $t_i > 0$ and $Et_i = \mu < \infty$. We think of t_i as the amount of time that our ith light bulb burns and let $T_n = t_1 + \ldots + t_n$ be the time that the nth light bulb burns out. Let $N_t = \min\{n : T_n \geq t\}$ be the number of light bulbs that we have used up to time t. Show that with probability one $N_t / t \to 1/\mu$. To get started, note that $\{N_t \leq ct\} = \{T_{ct} \geq t\}$, and apply the strong law of large numbers to T_{ct}.

1.15. Board games. Let $t_1, t_2, \ldots$ be i.i.d. uniform on $\{1, 2, \ldots, 6\}$ and think of t_i as the number of spaces you move on the ith roll in a board game. $N_t = \min\{n : T_n \geq t\}$ is the number of rolls you need to get to the finish if it is t spaces away. Use the result in the last exercise to show that $N_t / t \to 2/7$ as $t \to \infty$.

1.16. Records. Suppose $X_1, X_2, \ldots$ are independent and have a continuous distribution. We say a record occurs at time n and we set $Y_n = 1$ if $X_n > \max\{X_1, \ldots, X_{n-1}\}$, otherwise $Y_n = 0$. Let $R_n = Y_1 + \ldots + Y_n$ be the number of records that have been set by time n. In Chapter 4, we computed ER_n in Exercise 3.20 and $\text{var}(R_n)$ in Exercise 4.16. Use these two results to show that $R_n / \ln n \to 1$ in probability as $n \to \infty$.

1.17. Coupon collector's problem. Suppose we are trying to collect a complete set of n baseball cards. Suppose we buy them one at a time and each time we get a randomly chosen card. Let N_n be the number of cards we have to buy to get a complete set. In Chapter 4, we computed EN_n in Exercise 3.21 and $\text{var}(N_n)$ in Exercise 4.17. Use these two results to show that $N_n / (n \ln n) \to 1$ in probability as $n \to \infty$.

5.2. The Central Limit Theorem

The limit theorem in this section gets its name not only from the fact that it is of central importance but also because it shows that if you add up a large number of random variables with a fixed distribution with finite variance then, after a suitable scaling, the result is close to the normal distribution. To see why the normal plays a special role we note that in general if $X_1, \ldots, X_n$ are i.i.d. with mean μ and variance $0 < \sigma^2 < \infty$ then (1.3) and (1.4) imply that $\bar{X}_n$ has mean μ and variance σ^2/n, so $(\bar{X}_n - \mu)/(\sigma/\sqrt{n})$ has mean 0 and variance 1. However, if $X_1, \ldots, X_n$ are independent and normal(μ, σ^2) then since sums of independent normals are normal, $\bar{X}_n$ is normal$(\mu, \sigma^2/n)$ and $(\bar{X}_n - \mu)/(\sigma/\sqrt{n})$ is normal(0,1). This calculation shows that if the next result holds then the normal distribution must be the limit.

(2.1) **Central limit theorem.** Suppose $X_1, X_2, \ldots$ are i.i.d. and have $EX_i = \mu$ and var$(X_i) = \sigma^2$ with $0 < \sigma^2 < \infty$. Then as $n \to \infty$

$$P\left(\frac{\bar{X}_n - \mu}{\sigma/\sqrt{n}} \leq x\right) \to P(\chi \leq x)$$

where χ denotes a random variable with the standard normal distribution.

PROOF OF (2.1) ASSUMING $Ee^{tX} < \infty$ FOR $t \in (-t_0, t_0)$: The first step is to reduce to the case in which $\mu = 0$. Let $Y_k = X_k - \mu$. $EY_k = 0$, var$(Y_k) = $ var(X_k), and $\bar{Y}_n = \bar{X}_n - \mu$, so it suffices to show

$$P\left(\frac{\bar{Y}_n}{\sigma/\sqrt{n}} \leq x\right) \to P(\chi \leq x)$$

as $n \to \infty$.

To prove this we will compute the moment generating function of $\bar{Y}_n$ and show that it converges to the moment generating function of the normal distribution. Let $\phi(t) = Ee^{tY}$. (3.2) from Chapter 4 implies

$$E \exp(tY_k/\sigma\sqrt{n}) = \phi(t/\sigma\sqrt{n})$$

$Y_1, \ldots, Y_n$ are independent, so (3.8) from Chapter 4 implies that the m.g.f. of the sum is the product of the individual m.g.f.'s:

$$E \exp\left(\frac{t(Y_1 + \cdots + Y_n)}{\sigma\sqrt{n}}\right) = \phi(t/\sigma\sqrt{n})^n$$

Now $(Y_1 + \cdots + Y_n)/(\sigma\sqrt{n}) = \bar{Y}_n/(\sigma/\sqrt{n})$. Substituting this equality into the last equation and taking logarithms of both sides,

$$\ln\left\{E \exp\left(\frac{t\bar{Y}_n}{\sigma/\sqrt{n}}\right)\right\} = n \ln \phi(t/\sigma\sqrt{n})$$

To find the asymptotic behavior of the right-hand side we let $\psi(s) = \ln \phi(s)$ and compute the first two derivatives, using the facts that $\phi(0) = 1$, $\phi'(0) = EY = 0$, and $\phi''(0) = EY^2$.

$$\psi(s) = \ln \phi(s) \qquad\qquad \psi(0) = \ln \phi(0) = \ln 1 = 0$$

$$\psi'(s) = \frac{\phi'(s)}{\phi(s)} \qquad\qquad \psi'(0) = \frac{\phi'(0)}{\phi(0)} = \frac{EY}{1} = 0$$

$$\psi''(s) = \frac{\phi''(s)}{\phi(s)} - \frac{\phi'(s)^2}{\phi(s)^2} \qquad\qquad \psi''(0) = EY^2 - (EY)^2 = \sigma^2$$

L'Hôpital's rule implies that if f and g are twice differentiable functions with $f(0) = g(0) = 0$, $f'(0) = g'(0) = 0$, and $g''(0) \neq 0$ then

$$\lim_{x \to 0} \frac{f(x)}{g(x)} = \frac{f''(0)}{g''(0)}$$

Using this result with $f(x) = \psi(x)$, $g(x) = x^2$, and $x = t/\sigma\sqrt{n}$, we have

$$\lim_{n \to \infty} \frac{\ln \phi(t/\sigma\sqrt{n})}{(t/\sigma\sqrt{n})^2} = \frac{\sigma^2}{2}$$

$$\lim_{n \to \infty} n \ln \phi(t/\sigma\sqrt{n}) = \frac{t^2}{2}$$

$$\lim_{n \to \infty} \phi(t/\sigma\sqrt{n})^n = \lim_{n \to \infty} \exp\left(n \ln \phi(t/\sigma\sqrt{n})\right) = e^{t^2/2}$$

At this point we have shown that the m.g.f. of $\bar{Y}_n/(\sigma/\sqrt{n})$ converges to the m.g.f. of a standard normal. Since the m.g.f. determines the distribution (see (3.10) in Chapter 4) it should not be surprising that the last conclusion implies

$$P\left(\frac{\bar{Y}_n}{\sigma/\sqrt{n}} \leq x\right) \to P(\chi \leq x)$$

However, the rigorous justification of the last step (like that of (3.10) of Chapter 4) is beyond the scope of this book. $\square$

REMARK. The proof we have given uses a very strong assumption, $Ee^{tX} < \infty$ for $t \in (-t_0, t_0)$ with $t_0 > 0$, but it is not too far from the proof under the assumption $\text{var}(X) < \infty$. That proof follows the same outline but uses the characteristic function defined by

$$Ee^{itX} = E\cos(tX) + iE\sin(tX)$$

where i is the complex number $\sqrt{-1}$, instead of the m.g.f.

Example 2.1. Suppose you play roulette and bet \$1 on black each time. Your net winnings on the ith play, X_i, are 1 with probability $18/38$ and -1 with probability $20/38$. What is the probability your net winnings are ≥ 0 after 81 plays?

Before entering into the computations we invite the reader to look at Figure 5.2, which shows the net winnings of a player in a simulation of 900 plays of a game of roulette. Note that while there are fluctuations, there is a clear decreasing trend.

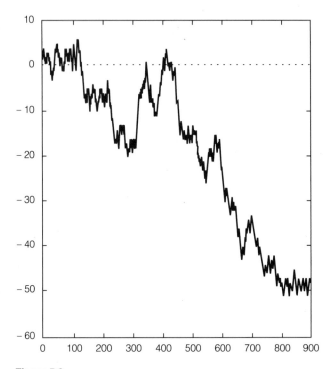

Figure 5.2

Computing the mean and variance of the X_i, we find

$$\mu = EX_i = 1(18/38) + (-1)20/38 = -1/19 = -0.05263$$
$$\sigma^2 = \text{var}(X_i) = EX_i^2 - (EX_i)^2 = 1 - (-1/19)^2 = 0.9972 \approx 1$$

where in the second computation $EX_i^2 = 1$ since X_i^2 is always 1, and we have approximated 0.9972 by 1 to make the arithmetic simpler. To write things so that we can apply the central limit theorem we note that

$$P(X_1 + \cdots + X_{81} \geq 0) = P(\bar{X}_{81} \geq 0) = P(\bar{X}_{81} - \mu \geq 1/19)$$

$$\approx P\left(\frac{\bar{X}_{81} - \mu}{\sigma/\sqrt{81}} \geq \frac{1/19}{1/9}\right)$$

$$\approx P(\chi \geq 9/19) = P(\chi \geq 0.47)$$

$$= 1 - P(\chi \leq 0.47) = 1 - 0.6808 = 0.3192$$

with the value of $P(\chi \leq 0.47)$ coming from the normal table in the Appendix.

Example 2.2. Suppose we flip a coin 900 times. What is the probability we get at least 465 heads?

Let $X_i = 1$ if we get heads on the ith toss and 0 otherwise so that $S_n = X_1 + \cdots + X_n$ is the number of heads in the first n tosses. To deal directly with S_n instead of translating things in terms of $\bar{X}_n$, we note that

$$\frac{\bar{X}_n - \mu}{\sigma/\sqrt{n}} = \frac{(S_n/n) - \mu}{\sigma/\sqrt{n}} = \frac{S_n - n\mu}{\sigma\sqrt{n}}$$

So the central limit theorem implies that if n is large,

$$(2.2) \qquad\qquad P\left(\frac{S_n - n\mu}{\sigma\sqrt{n}} \leq x\right) \approx P(\chi \leq x)$$

To use this result we note that the X_i have a Bernoulli distribution with $p = P(X_i = 1) = 1/2$, so $\mu = EX_i = p = 1/2$, $\sigma^2 = \text{var}(X_i) = p(1-p) = 1/4$, $\sigma = 1/2$, and

$$n\mu = 900 \cdot \frac{1}{2} = 450 \qquad \sigma\sqrt{n} = \frac{1}{2} \cdot \sqrt{900} = 15$$

Writing the probability of interest in such a way that we can use (2.2), we have

$$P(S_n \geq 465) = P\left(\frac{S_n - 450}{15} \geq \frac{15}{15}\right) \approx P(\chi \geq 1) = 1 - 0.8413 = 0.1587$$

If the question in the problem had been formulated as "What is the probability of at most 464 heads?" we would have computed

$$P(S_n \leq 464) = P\left(\frac{S_n - 450}{15} \leq \frac{14}{15}\right) \approx P(\chi \leq 0.933) = 0.8238$$

which does not quite agree with our first answer since

$$0.8238 + (1 - 0.8413) = 0.9825 < 1$$

whereas $P(S_n \leq 464) + P(S_n \geq 465) = 1$. The solution to this problem is to regard $\{S_n \geq 465\}$ as $\{S_n \geq 464.5\}$, that is, the integers 464 and 465 split up the territory that lies between them. When we do this, the answer to our original question becomes

$$P(S_n \geq 464.5) = P\left(\frac{S_n - 450}{15} \geq \frac{14.5}{15}\right)$$

$$\approx P(\chi \geq 0.966) = 1 - 0.8340 = 0.1660$$

which is a much better approximation of the exact probability 0.1669 than was our first answer, 0.1587.

The last correction, which is called the **histogram correction**, should be used whenever we apply (2.2) to integer-valued random variables. As we did in the last example, if k is an integer we regard $P(S_n \geq k)$ as $P(S_n \geq k - 0.5)$ and $P(S_n \leq k)$ as $P(S_n \leq k + 0.5)$. More generally, we replace each integer k in the set of interest by the interval $[k - 0.5, k + 0.5]$. The next example shows that the histogram correction is not only a device to get more accurate estimates, it also allows us to get answers in cases where a naive application of the central limit theorem would give a senseless answer.

Example 2.3. Suppose we perform 15 repetitions of an experiment with success probability 0.2. What is the probability of exactly 3 successes?

Let X_i be 1 if success occurs on the ith trial and 0 otherwise. The X_i have a Bernoulli distribution with $p = P(X_i = 1) = 0.2$, so $\mu = EX_i = p = 0.2$, $\sigma^2 = \text{var}(X_i) = p(1 - p) = 0.2(0.8) = 0.16$, and $\sigma = \sqrt{0.16} = 0.4$. Letting $S_n = X_1 + \cdots + X_n$ be the number of successes in $n = 15$ trials, we have

$$n\mu = 15(0.2) = 3 \qquad \sigma\sqrt{n} = 0.4\sqrt{15} = 1.549$$

If we used (2.2) without thinking, we would say that

$$P(S_n = 3) = P\left(\frac{S_n - 3}{1.549} = 0\right) \approx P(\chi = 0) = 0$$

but viewing $\{S_n = 3\}$ as $\{2.5 \leq S_n \leq 3.5\}$, we have

$$P(2.5 \leq S_n \leq 3.5) = P\left(\frac{-0.5}{1.549} \leq \frac{S_n - 3}{1.549} \leq \frac{0.5}{1.549}\right)$$

$$\approx P(-0.32 \leq \chi \leq 0.32) = 2P(0 \leq \chi \leq 0.32)$$

$$= 2(0.6255 - 0.5) = 0.2510$$

The same method can be applied to other values. The next table compares the results of those calculations with the exact probabilities computed from the binomial distribution. Figure 5.3 gives a graphical representation of the normal approximation in this case.

k	0	1	2	3	4	5	6
exact	.0352	.1319	.2309	.2501	.1876	.1032	.0430
normal	.0422	.1137	.2060	.2510	.2060	.1137	.0422
Poisson	.0498	.1494	.2240	.2240	.1680	.1008	.0504

As the last line of the table indicates, these probabilities can also be approximated by a Poisson(3) distribution.

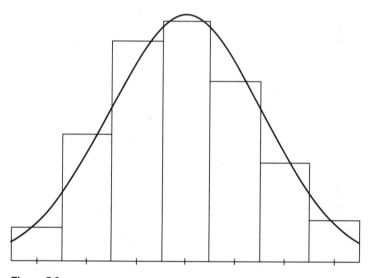

Figure 5.3

Since the binomial with parameter n and p can, when p is small and n is large, be approximated by a Poisson with parameter $\lambda = np$ or by a normal with mean np and variance $np(1-p)$, it should not be surprising that a Poisson with parameter λ can, when λ is large, be approximated by a normal with mean λ and variance λ.

Example 2.4. Suppose that the number of customers who arrive at a fast food restaurant between 11:30 and 1:30 has a Poisson distribution with mean 225. What is the probability that there will be fewer than 200 customers?

According to Example 7.2 in Chapter 3, if Y has a Poisson distribution with mean 225 then Y can be written as a sum of 225 independent Poisson(1) random variables X_i. Using the central limit theorem in the form (2.2) and recalling that

$$n\mu = ES_n = EY = \lambda$$
$$\sigma\sqrt{n} = \sqrt{\text{var}(S_n)} = \sqrt{\text{var}(Y)} = \sqrt{\lambda}$$

where $\lambda = 225$, we have

$$\chi \approx \frac{Y - EY}{\sqrt{\text{var}(Y)}} = \frac{Y - \lambda}{\sqrt{\lambda}} = \frac{Y - 225}{15}$$

Using the histogram correction and writing $\{Y < 200\}$ as $\{Y \le 199.5\}$, we have

$$P(Y \le 199.5) = P\left(\frac{Y - 225}{15} \le \frac{-25.5}{15}\right)$$
$$\approx P(\chi \le -1.7) = P(\chi \ge 1.7) = 1 - 0.9554 = 0.0446$$

The calculations in this example generalize easily to show that if $Y = \text{Poisson}(\lambda)$ and λ is large then

(2.3)
$$P\left(\frac{Y - \lambda}{\sqrt{\lambda}} \le x\right) \approx P(\chi \le x)$$

To prove this, let ℓ be the largest integer $\le \lambda$ and write Y as the sum of ℓ independent Poisson(λ/ℓ) random variables. The mean and variance of the sum are both λ since Y has mean λ and variance λ, so using (2.2) gives (2.3). To help the reader remember the approximations in (2.1)–(2.3) we note that in each case the approximation says

(2.4)
$$P\left(\frac{Z - EZ}{\sqrt{\text{var}(Z)}} \le x\right) \approx P(\chi \le x)$$

i.e., we subtract the mean and divide by the square root of the variance to make the mean 0 and the variance 1, and then the random variable is approximately standard normal. The formulas look different in the three cases since the formulas for the mean and variance are.

Z	EZ	$\text{var}(Z)$
$\bar{X}_n$	μ	σ^2/n
S_n	$n\mu$	$\sigma^2 n$
Poisson	λ	λ

To help remember where to put n in the first two cases, recall that when n is large $\bar{X}_n$ is close to μ, and hence has small variance, while the variance of S_n increases with n.

Exercises

2.1. In Example 1.3, we used Chebyshev's inequality to conclude that if we flip a coin 10,000 times then $P(|\bar{X}_n - 1/2| \geq 0.01) \leq 1/4$. Use the normal approximation to estimate this probability.

2.2. A person bets you that in 100 tosses of a fair coin the number of Heads will differ from 50 by 4 or more. What is the probability you will win this bet?

2.3. A basketball player makes 80% of his free throws on the average. Use the normal approximation to compute the probability that in 25 attempts he will make at least 23.

2.4. A student is taking a true/false test with 48 questions. (a) Suppose she has a probability $p = 3/4$ of getting each question right. What is the probability she will get at least 38 right? (b) Answer the last question if she knows the answers to half the questions and flips a coin to answer the other half. Notice that in each case the expected number of questions she gets right is 36.

2.5. Suppose 1% of all screws made by a machine are defective. We are interested in the probability a batch of 225 screws has at most one defective screw. Compute (a) the exact answer, (b) the Poisson approximation, (c) the normal approximation.

2.6. Forty-eight numbers are rounded off to the nearest integer and then added. If the individual round-off errors are uniformly distributed over $(-0.5, 0.5)$, what is the probability the sum of the rounded numbers differs from the true sum (a) by more than 3, (b) by more than 5? Notice that the most pessimistic estimate of the errors involved is ± 24.

2.7. Suppose we roll a die 200 times. What is the probability that the sum of the numbers obtained lies between 600 and 800?

2.8. Suppose the weight of a certain brand of bolt has a mean of 1 gram and a standard deviation of 0.13 grams. Estimate the probability that 100 of these bolts weigh more than 102 grams.

2.9. You have 12 problems to solve. Suppose that each solution requires an exponential amount of time with mean 1/2. What is the probability it will take more than 7 hours to solve all the problems?

2.10. Suppose $X = \text{gamma}(100, 2)$. Estimate $P(50 \le X \le 58)$.

2.11. Suppose $X = \text{Poisson}(100)$. Estimate $P(85 \le X \le 115)$.

2.12. Suppose that the checkout time at a grocery store has a mean of 5 minutes and a standard deviation of 2 minutes. Estimate the probability that a checker will serve more than 49 customers during her 4-hour shift.

2.13. A die is rolled repeatedly until the sum of the numbers obtained is larger than 200. What is the probability that you can do this in 66 rolls or fewer?

2.14. Renewal theory. Suppose $t_1, t_2, \ldots$ are i.i.d. with $t_i > 0$ and $Et_i = \mu < \infty$. We think of t_i as the amount of time that our ith lightbulb burns and $T_n = t_1 + \cdots + t_n$ as the time that the nth light bulb burns out. Let $N_t = \max\{n : T_n \le t\}$ be the number of light bulbs that have burned out by time t. Now

$$P(N_t \ge t/\mu + xt^{1/2}) = P(T_{t/\mu + xt^{1/2}} \le t)$$

Apply the central limit theorem to $T_{\mu t + xt^{1/2}}$ and conclude that

$$P\left(\frac{N_t - t/\mu}{t^{1/2}} \le x\right) \to P\left(\chi \le \frac{\mu^{3/2}}{\sigma} x\right)$$

where χ has the standard normal distribution.

2.15. Monte Carlo integration. Suppose r has $\int_0^1 r(x)^2\, dx < \infty$ and let $U_1, U_2, \ldots$ be i.i.d. uniform on $(0, 1)$. Let $I_n = (r(U_1) + \cdots + r(U_n))/n$, $I = \int_0^1 r(x)\, dx$, and $\sigma^2 = \int_0^1 r(x)^2\, dx - I^2$. Show that if $\sigma > 0$ then

$$P(\sigma^{-1} n^{1/2}(I_n - I) \le x) \to P(\chi \le x)$$

This is not a very fast rate of convergence since by evaluating f at n equally spaced points we have (if f is twice differentiable)

$$\frac{1}{n} \sum_{i=1}^{n} \frac{f(x_{i-1}) + f(x_i)}{2} - I = O(n^{-2})$$

However, Monte Carlo integration works even when the function is not smooth and has a convergence rate of $n^{-1/2}$ in any dimension. In contrast, with a regular grid in d dimensions, you need m^d points to achieve an accuracy of m^{-2}, or changing variables $n = m^d$, we have a convergence rate of $n^{-2/d}$.

5.3. Confidence Intervals

In this section and the next we will briefly discuss confidence intervals and hypothesis testing to illustrate the use of the central limit theorem in statistics. We will not attempt a systematic treatment since these topics are usually treated in detail in statistics courses that follow this one.

Suppose $X_1, \ldots, X_n$ is a random sample from a population with unknown mean μ but known variance σ^2. (That is, the X_i are i.i.d. with mean μ and variance σ^2.) It is a little strange to think that you could know the variance of the X_i but not know their mean, but this situation will help prepare us for the more realistic scenarios that follow it. Consulting the normal table, we find that

$$P(-2 \le \chi \le 2) = 2P(0 \le \chi \le 2) = 2(0.9772 - 0.5) = 0.9544$$

so using the central limit theorem, (2.1), we have

$$P\left(-\frac{2\sigma}{\sqrt{n}} \le \bar{X}_n - \mu \le \frac{2\sigma}{\sqrt{n}}\right) \approx 0.95$$

Rewriting the last result we have

(3.1)
$$P\left(\mu \in \left[\bar{X}_n - \frac{2\sigma}{\sqrt{n}}, \bar{X}_n + \frac{2\sigma}{\sqrt{n}}\right]\right) \approx 0.95$$

We say that $[\bar{X}_n - 2\sigma/\sqrt{n}, \bar{X}_n + 2\sigma/\sqrt{n}]$ is a **95% confidence interval** for μ since μ will lie in this interval with a probability that is approximately 0.95.

Example 3.1. Suppose we have a random sample of size 100 from a population with standard deviation $\sigma = 3$ and we observe a sample mean of $\bar{X}_{100} = 66.32$. (For a concrete example, think of measuring the height of 100 women.) In this case

$$\frac{2\sigma}{\sqrt{n}} = \frac{2 \cdot 3}{\sqrt{100}} = \frac{6}{10} = 0.6$$

so a 95% confidence interval for μ is $66.32 \pm 0.6 = [65.72, 66.92]$.

The 2 in the 95% confidence interval in (3.1) comes from setting $P(-2 \le \chi \le 2) = 0.9544$. If we do the arithmetic more carefully, we see that

$$P(-1.96 \le \chi \le 1.96) = 2P(0 \le \chi \le 1.96) = 2(0.9750 - 0.5) = 0.95$$

but we will stick with 2 since it makes the arithmetic much simpler. If for some reason you want a 90% confidence interval then you want to use

$$[\bar{X}_n - c\sigma/\sqrt{n}, \bar{X}_n + c\sigma/\sqrt{n}]$$

where c is chosen so that

$$P(-c \leq \chi \leq c) = 2P(0 \leq \chi \leq c) = 0.90$$

That is, $P(0 \leq \chi \leq c) = 0.45$, or $P(\chi \leq c) = 0.95$, and consulting the table we see $c = 1.65$. Penny pinchers use $c = 1.645$ based on the fact that

$$P(\chi \leq 1.64) = 0.9495$$
$$P(\chi \leq 1.65) = 0.9505$$

EXERCISE 3.1. What value of c should be taken to get a (a) 98% or (b) 99% confidence interval?

We turn now to the situation that occurs in most examples: We do not know the mean or the variance. The solution is simple – we use the sample to estimate σ^2. To motivate the estimate that we will use, we note that

$$\bar{X}_n = (X_1 + \cdots + X_n)/n$$

is the mean of the so-called **empirical distribution**, which assigns probability $1/n$ to each of the X_i. We use the mean of the empirical distribution $\bar{X}_n$ to estimate the mean μ, so it is natural to estimate the variance σ^2 by the variance of the empirical distribution:

$$s_n^2 = \frac{1}{n}\sum_{i=1}^{n}(X_i - \bar{X}_n)^2 = \left(\frac{1}{n}\sum_{i=1}^{n}X_i^2\right) - \bar{X}_n^2$$

Here the second equality follows from formula $E(Y - EY)^2 = EY^2 - (EY)^2$ applied to the random variable Y that is equal to X_i with probability $1/n$ and hence has $EY = \bar{X}_i$.

The first thing we want to show is

(3.2) With probability one, $s_n^2 \to \sigma^2$ as $n \to \infty$.

PROOF: The strong law of large numbers tells us that with probability one $\bar{X}_n \to \mu$, so $\bar{X}_n^2 \to \mu^2$ with probability one. Applying the strong law to the i.i.d. variables $Y_i = X_i^2$ we have

$$\frac{1}{n}\sum_{i=1}^{n}X_i^2 = \frac{1}{n}\sum_{i=1}^{n}Y_i = \bar{Y}_n \to EY = EX^2$$

with probability one. Combining this with the conclusion for $\bar{X}_n^2$ we have

$$s_n^2 = \left(\frac{1}{n} \sum_{i=1}^n X_i^2 \right) - \bar{X}_n^2 \to EX^2 - (EX)^2 = \text{var}(X)$$

with probability one. $\square$

Being the variance of the empirical distribution, our estimator of σ^2 is a natural one, but from another point of view it is not. s_n^2 is a **biased estimator**, that is, $Es_n^2 \neq \sigma^2$. To compute Es_n^2 we note that

$$ns_n^2 = \sum_{m=1}^n (X_m - \bar{X}_n)^2 = \sum_{m=1}^n (X_m - \mu - (\bar{X}_n - \mu))^2$$

$$= \sum_{m=1}^n \left\{ (X_m - \mu)^2 - 2(X_m - \mu)(\bar{X}_n - \mu) + (\bar{X}_n - \mu)^2 \right\}$$

$$= \sum_{m=1}^n (X_m - \mu)^2 - 2(\bar{X}_n - \mu) \sum_{m=1}^n (X_m - \mu) + n(\bar{X}_n - \mu)^2$$

$$= \sum_{m=1}^n (X_m - \mu)^2 - n(\bar{X}_n - \mu)^2$$

since $\sum_{m=1}^n (X_m - \mu) = n\bar{X}_n - n\mu$. Taking expected value, now we have

$$nEs_n^2 = \sum_{m=1}^n E(X_m - \mu)^2 - nE(\bar{X}_n - \mu)^2$$

$$= n\sigma^2 - n\text{var}(\bar{X}_n) = (n-1)\sigma^2$$

since $E\bar{X}_n = \mu$ and $\text{var}(\bar{X}_n) = \sigma^2/n$.

As a check on our calculation of Es_n^2 we note that when $n = 1$, $\bar{X}_1 = X_1$ so $s_1^2 = (X_1 - X_1)^2 = 0$, which agrees with our computation that $Es_1^2 = 0$. Since $Es_n^2 = \sigma^2(n-1)/n$, it follows that for $n > 1$, an unbiased estimator of σ^2 is

$$\frac{n}{n-1} s_n^2 = \frac{1}{n-1} \sum_{i=1}^n (X_i - \bar{X}_n)^2$$

Since we will only be interested in large samples we will stick with the biased but natural estimator s_n^2 and call it the **sample variance**.

Having settled on a way of estimating σ^2 there is very little more to say about 95% confidence intervals when σ^2 is unknown. We let $s_n = \sqrt{s_n^2}$ and replace σ by our estimate s_n to get our 95% confidence interval

(3.3)
$$\left[\bar{X}_n - \frac{2s_n}{\sqrt{n}}, \bar{X}_n + \frac{2s_n}{\sqrt{n}} \right]$$

To illustrate the use of the last formula we consider

Example 3.2. A random sample of size 100 generated on a computer produced the following result:

k	1	2	3	4	5	6	7	8	9	15
n_k	31	20	19	8	10	2	2	3	4	1

Here n_k gives the number of times k appeared in the sample. For example, 31 observations were 1; 20 were 2; 15 appeared only once, and the missing numbers not at all. The sample mean and variance can be computed by noting that

$$\bar{X}_{100} = \frac{X_1 + \cdots + X_{100}}{100} = \frac{1}{100}\sum_k k\,n_k = 3.11$$

$$\frac{1}{100}\sum_{i=1}^{100} X_i^2 = \frac{1}{100}\sum_k k^2\,n_k = 15.71$$

$$s_n^2 = 15.71 - (3.11)^2 = 6.038$$

So $s_n = 2.457$, $2s_n/\sqrt{n} = 4.914/10 = 0.49$, and the 95% confidence interval is

$$\bar{X}_n \pm \frac{2s_n}{\sqrt{n}} = 3.11 \pm 0.49 = [2.62, 3.6]$$

These answers are quite reasonable since the random variables generated by the computer were geometric with $p = 1/3$ and such a geometric has mean $1/p = 3$ and variance $(1 - p)/p^2 = (2/3)/(1/3)^2 = 6$.

An important special case of estimating a mean is estimating a proportion.

Example 3.3. Weldon's dice data. An English biologist named Weldon was interested in the "pip effect" in dice – the idea that the spots, or "pips," which on some dice are produced by cutting small holes in the surface, make the sides with more spots lighter and more likely to turn up. Weldon threw 12 dice 26,306 times for a total of 315,672 throws and observed that a 5 or 6 came up on 106,602 throws. His estimate of the probability that a 5 or 6 appears is thus $\hat{p} = 106,602/315,672 = 0.33770$. Now if the true probability is p then the variance of each observation is $\sigma^2 = p(1 - p)$ so we estimate σ by $\sqrt{\hat{p}(1 - \hat{p})}$ and arrive at a 95% confidence interval

(3.4)
$$\left[\hat{p} - \frac{2\sqrt{\hat{p}(1 - \hat{p})}}{\sqrt{n}}, \hat{p} + \frac{2\sqrt{\hat{p}(1 - \hat{p})}}{\sqrt{n}}\right]$$

Plugging in our estimate $\hat{p}$ we have

$$0.33770 \pm 2\sqrt{\frac{0.3377 \cdot 0.6623}{315,672}} = 0.33770 \pm 0.00168 = [0.33602, 0.33938]$$

so, for Weldon's dice at least, the true probability of a 5 or 6 is somewhat larger than 1/3. This difference is not enough to be noticeable by people who play dice games for amusement but is perhaps large enough to be of concern for a casino that entertains tens of thousands of gamblers a year. For this reason most casinos use dice with no pips.

Example 3.4. Sample size selection. Suppose you want to forecast the outcome of an election and you are trying to figure out how many people to survey. From (3.4) we see that if a fraction $\hat{p}$ of the people in a sample of size n are for candidate B then the 95% confidence interval will be

$$\hat{p} \pm \frac{2\sqrt{\hat{p}(1 - \hat{p})}}{\sqrt{n}}$$

To get rid of the $\hat{p}$'s in the width of the confidence interval we note that the function $g(x) = x(1-x) = x - x^2$ has derivative $g'(x) = 1 - 2x$, which is positive for $x < 1/2$ and negative for $x > 1/2$. So g is increasing for $x < 1/2$, decreasing for $x > 1/2$, and hence the maximum value occurs at $x = 1/2$. Noticing that $2\sqrt{x(1 - x)} = 1$ when $x = 1/2$, we have

(3.5) $$P\left(p \in \left[\hat{p} - \frac{1}{\sqrt{n}}, \hat{p} + \frac{1}{\sqrt{n}}\right]\right) \geq 0.95$$

To see that this approximation is reasonable for elections, notice that if $0.4 \leq p \leq 0.6$ then $\sqrt{p(1 - p)} \geq \sqrt{0.24} = 0.4899$ compared with our upper bound of 1/2. Even when p is as small as 0.2, $\sqrt{p(1 - p)} = \sqrt{0.16} = 0.4$.

Suppose now that we want to estimate the outcome of the election within 2%. Setting $1/\sqrt{n} = 0.02$ and solving gives $n = (1/0.02)^2 = 50^2 = 2500$. To get the error down to 1% we would need $n = 1/(0.01)^2 = 10,000$. Comparing the last two results and noticing that the radius of the confidence interval in (3.5) is $1/\sqrt{n}$, we see that to reduce the error by a factor of 2 requires a sample that is $2^2 = 4$ times as large.

Example 3.5. The Literary Digest poll. In order to forecast the outcome of the 1936 election, *Literary Digest* polled 2.4 million people and found that 57% of them were going to vote for Alf Landon and 43% were going to vote for F. D. Roosevelt. A 95% confidence interval for the true fraction of people voting for Landon based on this sample would be 0.57 ± 0.00064 but Roosevelt won, getting 62% of the vote to Landon's 38%. To explain how this happened we have to look at the methods *Literary Digest* used. They sent 10 million questionnaires to people whose names came from telephone books and club membership lists. Since many of the 9 million unemployed did not belong to clubs or have telephones the sample was not representative of the population as a whole. A second bias came from the fact that only 24% of the people filled out the form. This problem was mentioned in our discussion of exit polls in Example 4.1 in Chapter 2. If, for example, 36% of Landon voters and 16.6% of Roosevelt voters responded then the fraction of people who responded would be $0.62(0.166) + 0.38(0.36) = 0.24$ and the fraction in the sample for Landon would be

$$\frac{0.38(0.36)}{0.62(0.166) + 0.38(0.36)} = \frac{0.1368}{0.24} = 0.57$$

in agreement with the data.

Finally, we would like to observe that *Literary Digest*, which soon after went bankrupt, could have saved a lot of money by taking a smaller sample. George Gallup, who was just then getting started in the polling business, predicted based on a sample of size 50,000 that Roosevelt would get 56% of the vote. His 95% confidence interval for the election result would be 0.56 ± 0.0045, compared with the election result of 62%. Again there could be some bias in his sample, or perhaps Landon voters, discouraged by the predicted outcome, were less likely to vote. The moral of our story is: It is much better to take a good sample than a large one.

EXERCISES

3.2. Suppose the time to complete a job has a normal distribution with a mean of 14 days and a standard deviation of 2 days. Find an interval so that the completion time of the job will be in there 75% of the time.

3.3. A light bulb has a brightness measured in footcandles that has a normal distribution with a mean of 2500 and a standard deviation of 70. Find a lower bound for the brightness that 98% of the light bulbs will exceed.

3.4. Inspection of 25 trout revealed a sample mean of 2.10 and a standard deviation of 0.25 pounds. Find a 95% confidence interval for the mean.

3.5. A machine measures the bounce of 36 tennis balls and finds a mean of 1.7 ft and a standard deviation of 0.3 ft. What is a 95% confidence interval for the bounce of tennis balls?

3.6. The mean length of time 64 butterflies spent in the pupa stage was 49 hours with a standard deviation of 10 hours. Find a 95% confidence interval for the mean duration of the pupa stage.

3.7. The tread life of 25 tires was an average of 31,485 miles with a standard deviation of 5,120 miles. Find a 95% confidence interval for the mean tread life.

3.8. Of the first 10,000 votes cast in an election, 5,180 were for candidate A. Find a 95% confidence interval for the fraction of votes that candidate A will receive.

3.9. Suppose we take a poll of 2,500 people. What percentage should the leader have for us to be 99% confident that the leader will be the winner?

3.10. A bank examines the records of 150 patrons and finds that 63 have savings accounts. Find a 95% confidence interval for the fraction of people with savings accounts.

3.11. Among 625 randomly chosen Swedish citizens, it was found that 25 had previously been citizens of another country. Find a 95% confidence interval for the true proportion.

3.12. Suppose $X_1, \ldots, X_{400}$ are independent and exponential(λ) for some unknown value of λ. If $\bar{X}_{400} = 3$, find a 95% confidence interval for $\mu = 1/\lambda$.

3.13. Suppose light bulbs have an exponential lifetime. Find a 95% confidence interval for the mean if the average lifetime of 25 bulbs is 1000 hours.

3.14. Suppose that the number of robberies that happen in a large city has a Poisson distribution. Find a 95% confidence interval for the number of robberies if the average number for the last nine years is 324.

3.15. Suppose we ask n people how they are going to vote in a presidential election and find that no one in our sample is going to vote for the Socialist Party candidate. It is not reasonable to use our formula to conclude that the single point $\{0\}$ is a 95% confidence interval. To get a sensible answer, let $h(p)$ be the probability of getting no Socialist voters in our sample when the true fraction is p, let $p_0 = \max\{p : h(p) < 0.05\}$, and let $[0, p_0]$ be our 95% confidence interval. (a) Find a formula for p_0. (b) Find p_0 when $n = 900$.

3.16. Suppose $X_1, \ldots, X_n$ are independent and uniform on $(0, \theta)$ where θ is not known. It is clear that

$$\theta \geq M_n = \max_{1 \leq i \leq n} X_i$$

If we let $p(\theta, x)$ denote the probability $M_n \leq x$ when the true parameter is θ then there is a less than 5% probability θ is larger than $N_n = \min\{\theta : p(\theta, M_n) \leq 0.05\}$. So $[M_n, N_n]$ is a 95% confidence interval for θ. (a) Find a formula for N_n. (b) Find N_n when $n = 100$ and $M_n = 3$.

5.4. Hypothesis Testing

There are many varieties of hypothesis tests but we will restrict our attention to two: testing a mean and testing the difference between two means.

Example 4.1. Suppose we run a casino and we wonder if our roulette wheel is biased. To rephrase our question in statistical terms, let p be the probability red comes up and introduce two hypotheses:

$$H_0 : p = 18/38 \qquad \text{null hypothesis}$$
$$H_1 : p \neq 18/38 \qquad \text{alternative hypothesis}$$

To test to see if the null hypothesis is true, we spin the roulette wheel n times and let $X_i = 1$ if red comes up on the ith trial and 0 otherwise, so that $\bar{X}_n$ is the fraction of times red came up in the first n trials. The test is specified by giving a **critical region** C_n so that we reject H_0 (that is, decide H_0 is incorrect) when $\bar{X}_n \in C_n$. One possible choice in this case is

$$C_n = \left\{ x : \left| x - \frac{18}{38} \right| > 2 \sqrt{\frac{18}{38} \cdot \frac{20}{38}} \bigg/ \sqrt{n} \right\}$$

This choice is motivated by the fact that if H_0 is true then using the central limit theorem,

(4.1) $\qquad P(\bar{X}_n \in C_n) = P\left(\left| \frac{\bar{X}_n - \mu}{\sigma/\sqrt{n}} \right| \geq 2 \right) \approx P(|\chi| \geq 2) = 0.05$

Rejecting H_0 when it is true is called a **type I error**. In this test we have set the type I error to be 5%.

Now $\sqrt{\frac{18}{38} \cdot \frac{20}{38}} = 0.4993 \approx 1/2$, so to simplify the arithmetic the test can be formulated as

$$\text{reject } H_0 \text{ if } \left| \bar{X}_n - \frac{18}{38} \right| > \frac{1}{\sqrt{n}}$$

or in terms of the total number of reds, $S_n = X_1 + \cdots + X_n$,

$$\text{reject } H_0 \text{ if } \left| S_n - \frac{18n}{38} \right| > \sqrt{n}$$

Suppose now that we spin the wheel $n = 3800$ times and get red 1868 times. Is the wheel biased? We expect $18n/38 = 1800$ reds, so the excess number of reds $|S_n - 1800| = 68$. Given the large number of trials, this might not seem like a large excess. However, $\sqrt{3800} = 61.6$ and $68 > 61.6$ so we reject H_0 and think,

"If H_0 were correct then we would see an observation this far from the mean less than 5% of the time."

Example 4.2. Do married college students with children do less well because they have less time to study or do they do better because they are more serious? These questions were running through the mind of a dean at a mythical university where the average GPA for all students is 2.48, so he decided on the following hypothesis test concerning μ, the mean grade point average of married students with children:

$$H_0 : \mu = 2.48 \qquad \text{null hypothesis}$$
$$H_1 : \mu \neq 2.48 \qquad \text{alternative hypothesis}$$

To test this hypothesis he consulted the records of 25 married college students with children and found an average GPA of 2.35 and a standard deviation of 0.5. Using (4.1) from the last example, we see that to have a test with a type I error of 5% he should reject H_0 if

$$(4.2) \qquad\qquad |\bar{X}_n - 2.48| > \frac{2\sigma}{\sqrt{n}}$$

Taking $n = 25$ and using the sample standard deviation to estimate σ we see that

$$\frac{2(0.5)}{\sqrt{25}} = 0.2 > 0.13 = |\bar{X}_n - 2.48|$$

so we are not 95% certain that $\mu \neq 2.48$.

Example 4.3. Testing the difference of two means. Suppose we have independent random samples of size n_1 and n_2 from two populations with unknown means μ_1, μ_2 and variances σ_1^2, σ_2^2 and we want to test

$$H_0 : \mu_1 = \mu_2 \qquad \text{null hypothesis}$$
$$H_1 : \mu_1 \neq \mu_2 \qquad \text{alternative hypothesis}$$

Now the central limit theorem implies that

$$\bar{X}_1 \approx \text{normal}\left(\mu_1, \frac{\sigma_1^2}{n_1}\right) \qquad -\bar{X}_2 \approx \text{normal}\left(-\mu_2, \frac{\sigma_2^2}{n_1}\right)$$

and we have assumed that the two samples are independent, so if H_0 is correct,

$$\bar{X}_1 - \bar{X}_2 \approx \text{normal}\left(0, \frac{\sigma_1^2}{n_1} + \frac{\sigma_2^2}{n_2}\right)$$

Based on the last result, if we want a test with a type I error of 5% then we should

$$(4.3) \qquad \text{reject } H_0 \text{ if } |\bar{X}_1 - \bar{X}_2| > 2\sqrt{\frac{\sigma_1^2}{n_1} + \frac{\sigma_2^2}{n_2}}$$

For a concrete example we consider a study of passive smoking reported in the *New England Journal of Medicine*, vol. 320 (1980), 720–723. A measurement of the size of lung airways called "FEF 25–75%" was taken for 200 female nonsmokers who were in a smoky environment and for 200 who were not. In the first group the average value of this statistic was 2.72 with a standard deviation of 0.71, while in the second group the average was 3.17 with a standard deviation of 0.74. (Larger values are better.) To see that there is a significant difference between the averages we note that

$$2\sqrt{\frac{\sigma_1^2}{n_1} + \frac{\sigma_2^2}{n_2}} = 2\sqrt{\frac{(0.71)^2}{200} + \frac{(0.74)^2}{200}} = 0.14503$$

while $|\bar{X}_1 - \bar{X}_2| = 0.45$.

Example 4.4. Paired comparisons. In our analysis of the last example, it was important that the two random samples were independent. If, for example, we use the same set of subjects to complete two tasks and compare their performance this assumption is not satisfied. For a concrete example, suppose that a computer scientist is investigating the ease of use of two different programming languages. She asks 25 experienced programmers to code an algorithm in each language and records their times. In this case we are interested in testing the hypothesis that the mean times to complete the task using the two languages are equal but the two samples are not independent, so we take the differences between the two times to get one sequence of independent observations and do the following hypothesis test about the mean of the difference μ:

$$H_0 : \mu = 0 \qquad \text{null hypothesis}$$
$$H_1 : \mu \neq 0 \qquad \text{alternative hypothesis}$$

To have a test with a type I error of 5% she should

$$(4.4) \qquad \text{reject } H_0 \text{ if } |\bar{X}_n| > \frac{2\sigma}{\sqrt{n}}$$

Suppose now that she found the average of the differences was 2.12 minutes with a standard deviation of 4.26. In this case,

$$\frac{2 \cdot 4.26}{\sqrt{25}} = 1.704 < 2.12 = \bar{X}_{25}$$

and we reject H_0. It seems that the second language is easier to use.

One of the reasons for doing a paired comparison is to reduce the variance. Suppose instead she had two sets of 25 programmers and found that the first set took 27.81 minutes with a standard deviation of 4.26 while the second took 25.69 minutes with a standard deviation of 4.26. The difference is still 2.12 minutes, but this time the difference is not significant since she must compare 2.12 with

$$2\sqrt{2\frac{(4.26)^2}{25}} = \sqrt{2} \cdot 1.704 = 2.410$$

Example 4.5. Our final example concerns the difference between two proportions, and its conclusion is paradoxical. When we consider men or women separately, a new treatment has a significant beneficial effect, but when the data on men and women is combined the new treatment seems to make people significantly worse! In the tables below, the two columns give results for people with and without the treatment.

MEN	with	without
recovered	670	80
not recovered	780	130

WOMEN	with	without
recovered	144	400
not recovered	71	280

COMBINED	with	without
recovered	814	480
not recovered	851	410

In each case, if we let p_1 be the probability of recovering with treatment and p_2 be the probability of recovering without, we want to test

$$H_0 : p_1 = p_2 \qquad \text{null hypothesis}$$
$$H_1 : p_1 \neq p_2 \qquad \text{alternative hypothesis}$$

Writing $\hat{p}_i$ for our estimate of p_i and recalling that the variance of a Bernoulli is $p(1-p)$, one might guess by analogy with the calculations in Example 4.3 that we should

$$\text{reject } H_0 \text{ if } |\hat{p}_1 - \hat{p}_2| > 2\sigma_n \quad \text{where} \quad \sigma_n = \sqrt{\frac{\hat{p}_1(1-\hat{p}_1)}{n_1} + \frac{\hat{p}_2(1-\hat{p}_2)}{n_2}}$$

However, if the null hypothesis is true, $p_1 = p_2$ and it makes sense to combine the two samples to get a better estimate of p and hence of the variance $p(1-p)$.

To do this we note that if there are n_i observations of type i then the fraction of successes in the entire sample is

$$\hat{p} = \frac{n_1\hat{p}_1 + n_2\hat{p}_2}{n_1 + n_2}$$

and we reject H_0 if

(4.5) $|\hat{p}_1 - \hat{p}_2| > 2\sigma_n$ where $\sigma_n = \sqrt{\hat{p}(1-\hat{p})\left(\dfrac{1}{n_1} + \dfrac{1}{n_2}\right)}$

Plugging in the data from the table, we find

	$\hat{p}_1$	$\hat{p}_2$	$\hat{p}_1 - \hat{p}_2$	$2\sigma_n$
MEN	.4621	.3810	.0811	.0735
WOMEN	.6698	.5882	.0815	.0764
COMBINED	.4889	.5393	−.0504	.0415

so in each case we reject H_0, but in the first two cases because $\hat{p}_1 - \hat{p}_2 > 2\sigma_n$ and in the last because $\hat{p}_1 - \hat{p}_2 < -2\sigma_n$. The problem here lies not in the hypothesis test but in the data itself:

$$\frac{670}{1450} > \frac{80}{210} \quad \text{and} \quad \frac{144}{215} > \frac{400}{680} \quad \text{but} \quad \frac{670 + 144}{1450 + 215} < \frac{80 + 400}{210 + 680}$$

EXERCISES

In the hypothesis testing problems below you are supposed to set up a test with a type I error of 5% and see if H_0 should be rejected.

4.1. A casino owner is concerned based on past experience that his dice show 6 too often so he decides to test the hypothesis $H_0 : p = 1/6$ against the alternative $H_1 : p \neq 1/6$. He makes his employees roll a die 18,000 times and they observe 3,123 sixes. Is the die biased?

4.2. We suspect that a bridge player is cheating by putting an honor card (Ace, King, Queen, or Jack) at the bottom of the deck when he shuffles so that this card will end up in his hand. Thus we let p be the probability that the last card dealt to him is an honor and we set out to test the hypothesis $H_0 : p = 4/13$ against the alternative $H_1 : p \neq 4/13$. In 16 times when he dealt, the last card dealt to him was an honor on 9 occasions. Are we 95% certain he is cheating?

4.3. A semiconductor firm makes computer chips. The contract with their customer calls for no more than 2% defective chips. If each shipment consists of 400 chips then how many defectives must there be to reject the hypothesis $H_0 : p = 0.02$ in favor of the alternative $H_1 : p \neq 0.02$?

4.4. If both parents carry one dominant (A) and one recessive gene (a) for a trait then Mendelian inheritance predicts that $1/4$ of the offspring will have both recessive genes (aa) and show the recessive trait. If among 96 offspring of Aa parents we find 30 are aa, is this consistent with Mendelian inheritance?

4.5. The mean weight of college men is 165 pounds with a standard deviation of 15 pounds. 25 students in an ROTC class have an average weight of 160.5. Are they significantly lighter than average?

4.6. The Intelligence Quotient (or IQ for short) has by definition a mean of 100 and a standard deviation of 20. 36 students in the senior class at a private school have an average IQ of 106. Are they significantly smarter than the average?

4.7. A statistically minded calculus professor observes that the 216 students who almost always come to class scored an average of 74 with a standard deviation of 12 while the 75 students who almost never come to class scored an average of 71.5 with a standard deviation of 10. Use this data to test the hypothesis $H_0 : \mu_1 = \mu_2$ against the alternative $H_1 : \mu_1 \neq \mu_2$.

4.8. The makers of the ABZ reading plan want to demonstrate that their method of teaching reading produces superior results. A group of 25 students taught by the ABZ method scored an average of 104 points on a reading test with a standard deviation of 15, while 16 students taught by traditional methods scored an average of 92 points with a standard deviation of 16. Are we 95% confident that $\mu_1 \neq \mu_2$?

4.9. Two fraternities had an ice cream eating contest. 27 guys from fraternity α ate an average of 3 bowls with a standard deviation of 3, while 24 guys from fraternity β ate an average of 6.5 bowls with a standard deviation of 4. Are we 95% confident that there is a difference between the two?

4.10. Suppose we take a sample of $n_1 = 400$ families in Philadelphia and find an average income of $\bar{X}_1 = 21K$, $s_1 = 16K$ (where K stands for thousands of dollars) while a similar study of $n_2 = 250$ families in Washington, D.C., finds $\bar{X}_2 = 18K$, $s_2 = 15K$. Is there a significant difference between the average incomes in these two cities?

4.11. A potential buyer of light bulbs bought 50 bulbs of each of two brands. On testing these bulbs, she found that brand A had a mean life of 1282 hours while brand B had a mean life of 1227 hours. If we assume that light bulb lifetimes have a standard deviation of 200 hours, are we 95% confident that $\mu_A \neq \mu_B$?

4.12. In the school district of a large city there were only 656 assaults this year, compared with an average of 680 in the previous five years. Assuming that the number of assaults has a Poisson distribution, are we 95% certain that the mean number of assaults has decreased?

4.13. In a poll of 500 people, 46% of 100 women and 51% of 400 men were in favor of a candidate. Is there a real difference in his popularity with men and women?

4.14. Two types of computers to control the firing of 105 mm weapons are being considered by the U.S. Army. Computer system 1 gave 121 hits in 150 attempts, while system 2 hit 104 times in 150 attempts. Is there a real difference between the two systems?

4.15. Two machines are used to fill bottles with dishwashing liquid. Examination of 12 bottles filled by machine 1 revealed a mean of 31.08 ounces with a standard deviation of 0.18, while 10 bottles filled by machine 2 had a mean of 30.62 with a standard deviation of 0.25. Find a 95% confidence interval for the difference of the two means.

4.16. The output voltage of two brands of transformers was tested. A sample of 10 of Brand A had a mean of 12.13 and a standard deviation of 0.8, while 12 of Brand B had a mean of 11.87 and a standard deviation of 0.7. Find a 95% confidence interval for the difference of the means.

4.17. You are planning a taste test with your brand of soft drink and another one. How large a sample would you need to be able to detect a difference if only 55% of the people prefer your soft drink?

4.18. Suppose that we wish to test the hypothesis $H_0 : \mu_1 = \mu_2$ against the alternative $H_1 : \mu_1 \neq \mu_2$ when the standard deviations are known to be $\sigma_1 = 3$ and $\sigma_2 = 2$. If we are going to use a fixed total sample size N, what fraction of the sample should be allocated to the first type to minimize the variance of $\bar{X}_1 - \bar{X}_2$?

4.19. An engineer wants the melting point of an alloy to be 1000 degrees and if it differs by more than 5 degrees he must change its composition. Suppose the standard deviation of each melting point measurement is 10 degrees. How many samples should he test so that with probability 0.95 he will reject the hypothesis $H_0 : \mu = 1000$ in favor of the alternative $H_1 : \mu \neq 1000$ when the true mean is 995?

4.20. The point of this exercise is to show that if we flip a fair coin long enough we will decide that it is biased. Let S_n be the number of heads at time n and let G_n, for "good at time n," be the event $|S_n - n/2| \leq \sqrt{n}$, i.e., S_n is within two standard deviations of its mean. Use the central limit theorem to show that there is a $\delta > 0$ so that $P(G_{4n}^c|G_n) \geq \delta > 0$ for all n, and then improve the last conclusion a little to conclude that if $E_k = G_{4^k}$ then $P(E_1 \cap \ldots \cap E_k) \leq (1 - \delta)^{k-1}$.

5.5. Chapter Summary and Review Problems

Section 5.1. Laws of large numbers. Suppose $X_1, X_2, \ldots$ are i.i.d., that is, are independent and identically distributed, and have $EX_i = \mu$. Let

$$\bar{X}_n = \frac{X_1 + \cdots + X_n}{n}$$

be the **sample mean**. The **weak law of large numbers** says that

(1.6) As $n \to \infty$, $\bar{X}_n \to \mu$ in probability, that is, for any $\epsilon > 0$,

$$P(|X_n - \mu| > \epsilon) \to 0$$

The **strong law of large numbers** says

(1.7) With probability one, $\bar{X}_n \to \mu$ as $n \to \infty$.

Section 5.2. The central limit theorem. Suppose $X_1, X_2, \ldots$ are i.i.d. with $EX_i = \mu$ and $\text{var}(X_i) = \sigma^2 \in (0, \infty)$. We have calculated (see (1.3) and (1.4)) that

$$E\bar{X}_n = \mu \quad \text{and} \quad \text{var}(\bar{X}_n) = \sigma^2/n$$

The **central limit theorem** says that as $n \to \infty$,

(2.1)
$$P\left(\frac{\bar{X}_n - \mu}{\sigma/\sqrt{n}} \leq x\right) \to P(\chi \leq x)$$

where χ has the standard normal distribution. For some purposes it is easier to write the last conclusion in terms of $S_n = X_1 + \cdots + X_n = n\bar{X}_n$. As $n \to \infty$,

(2.2)
$$P\left(\frac{S_n - n\mu}{\sigma\sqrt{n}} \leq x\right) \to P(\chi \leq x)$$

A typical application of (2.2) is to approximate the binomial distribution by a normal distribution. When we approximate an integer-valued S_n we use the **histogram correction**. That is, we view inequalities of the form $S_n \geq k$ as $S_n \geq k - 0.5$ or, more generally, replace each integer k in the set of interest by the interval $[k - 0.5, k + 0.5]$.

Let $Y_\lambda = \text{Poisson}(\lambda)$ and ℓ be the largest integer $\leq \lambda$. Since Y_λ is the sum of ℓ independent $\text{Poisson}(\lambda/\ell)$ random variables, (2.2) implies that as $\lambda \to \infty$,

$$(2.3) \qquad P\left(\frac{Y_\lambda - \lambda}{\sqrt{\lambda}} \leq x\right) \to P(\chi \leq x)$$

To see that the three results in (2.1), (2.2), and (2.3) are closely related, note that in each case we take a random variable Z, subtract its mean EZ, and divide by $\sqrt{\text{var}(Z)}$ to get a random variable with mean 0 and variance 1, which the central limit theorem tells us has approximately a standard normal distribution.

Section 5.3. Confidence intervals. Using the fact that $P(-2 \leq \chi \leq 2) = 0.9544$ in the central limit theorem we see that ·

$$(3.1) \qquad P\left(\mu \in \left[\bar{X}_n - \frac{2\sigma}{\sqrt{n}}, \bar{X}_n + \frac{2\sigma}{\sqrt{n}}\right]\right) \approx 0.95$$

so if σ^2 is known $[\bar{X}_n - 2\sigma/\sqrt{n}, \bar{X}_n + 2\sigma/\sqrt{n}]$ gives a 95% confidence interval for μ. If we don't know σ then we define the **sample variance** by

$$s_n^2 = \frac{1}{n}\sum_{i=1}^{n}(X_i - \bar{X}_n)^2 = \left(\frac{1}{n}\sum_{i=1}^{n}X_i^2\right) - \bar{X}_n^2$$

and use $s_n = \sqrt{s_n^2}$ to estimate σ. Our 95% confidence interval in this case is

$$(3.3) \qquad \left[\bar{X}_n - \frac{2s_n}{\sqrt{n}}, \bar{X}_n + \frac{2s_n}{\sqrt{n}}\right]$$

If we are estimating a proportion (i.e., $X_i = 1$ with probability p and 0 with probability $1 - p$) we write $\hat{p}$ instead of $\bar{X}_n$ and estimate σ^2 by $\hat{p}(1 - \hat{p})$, so our confidence interval is

$$(3.4) \qquad \left[\hat{p} - \frac{2\sqrt{\hat{p}(1-\hat{p})}}{\sqrt{n}}, \hat{p} + \frac{2\sqrt{\hat{p}(1-\hat{p})}}{\sqrt{n}}\right]$$

If we want to take a poll and are planning our sample size, it is useful to observe that $2\sqrt{p(1-p)} \leq 1$ so the radius of our confidence interval will always be smaller than $1/\sqrt{n}$.

Section 5.4. Hypothesis testing. Suppose we want to test

$$H_0 : \mu = \mu_0 \qquad \text{null hypothesis}$$
$$H_1 : \mu \neq \mu_0 \qquad \text{alternative hypothesis}$$

where μ_0 is a given number, with a type I error of 5%. That is, we want the probability that we decide that H_0 is false when it is actually true to be 5%. then we reject H_0, i.e., decide it is not correct, if

$$|\bar{X}_n - \mu_0| > \frac{2s_n}{\sqrt{n}}$$

Suppose we are considering a proportion (that is, $X_i = 1$ with probability p and 0 with probability $1 - p$) and performing the test

$$H_0 : p = p_0 \qquad \text{null hypothesis}$$
$$H_1 : p \neq p_0 \qquad \text{alternative hypothesis}$$

where p_0 is a given number. If H_0 is correct $\sigma = \sqrt{p_0(1 - p_0)}$, so we reject H_0 if our estimate of p, $\hat{p}$, satisfies

$$|\hat{p} - p_0| > \frac{2\sqrt{p_0(1 - p_0)}}{\sqrt{n}}$$

Suppose we want to test

$$H_0 : \mu_1 = \mu_2 \qquad \text{null hypothesis}$$
$$H_1 : \mu_1 \neq \mu_2 \qquad \text{alternative hypothesis}$$

with a type I error of 5% on the basis of two independent samples of size n_1 and n_2 with sample means $\bar{X}_1$ and $\bar{X}_2$. We reject H_0 if

$$|\bar{X}_1 - \bar{X}_2| > 2\sqrt{\frac{s_1^2}{n_1} + \frac{s_2^2}{n_2}}$$

where s_1 and s_2 are the sample standard deviations.

Finally, suppose we are looking at the difference of two proportions

$$H_0 : p_1 = p_2 \qquad \text{null hypothesis}$$
$$H_1 : p_1 \neq p_2 \qquad \text{alternative hypothesis}$$

on the basis of two independent samples of size n_1 and n_2 with sample proportions $\hat{p}_1$ and $\hat{p}_2$. If H_0 is correct then $p_1 = p_2$, so to estimate the variance we combine the samples to get our best estimate of p,

$$\hat{p} = \frac{n_1 \hat{p}_1 + n_2 \hat{p}_2}{n_1 + n_2}$$

and then we reject H_0 if

$$|\hat{p}_1 - \hat{p}_2| > 2\sqrt{\hat{p}(1 - \hat{p})\left(\frac{1}{n_1} + \frac{1}{n_2}\right)}$$

In each of the four hypothesis testing situations we reject H_0 if our test statistic differs from the value prescribed by H_0 by more than 2 standard deviations. The tests look different only because the formulas for the standard deviations are.

EXERCISES

5.1. Suppose we toss a coin 100 times. Which is bigger, the probability of exactly 50 Heads or at least 60 Heads?

5.2. Suppose that 10% of a certain brand of jelly beans are red. Use the normal approximation to estimate the probability that in a bag of 400 jelly beans there are at least 45 red ones.

5.3. A softball player brags that he is a .300 hitter, yet at the end of the season he has gotten 21 hits in 84 at bats. Is this just bad luck? To decide, compute the probability that he would get 21 hits or less if his probability of getting a hit were $p = 0.3$.

5.4. Suppose that 15% of people don't show up for a flight, and suppose that their decisions are independent. How many tickets can you sell for a plane with 144 seats and be 99% sure that not too many people will show up?

5.5. A gymnast has a difficult trick with a 10% chance of success. She tries the trick 25 times and wants to know the probability she will get exactly two successes. Compute the (a) exact answer, (b) Poisson approximation, (c) normal approximation.

5.6. An airplane contains 36 men whose weights are normally distributed with mean 175 and standard deviation 16 pounds. (a) Estimate the probability that their total weight will be more than 6,500 pounds. (b) Find a number w_0 so that with probability 0.95 the sum of the weights will be less than w_0.

5.7. Members of the Beta Upsilon Tau fraternity each drink a random number of beers with mean 6 and standard deviation 3. If there are 81 fraternity members, how much should they buy so that using the normal approximation they are 93.32% sure they will not run out?

5.8. A test of a medicine for reducing blood pressure revealed that 16 patients had a mean reduction of 11 points in their diastolic blood pressure (the smaller number in the reading) with a standard deviation of 2 points. (a) Use the normal approximation to find a 95% confidence interval for the mean reduction. (b) Use Chebyshev's inequality to find an interval that we know the mean will lie in with a probability of at least 95%.

5.9. A sample of 2,809 hand-held video games revealed that 212 broke within the first three months of operation. Find a 95% confidence interval for the true proportion that break in the first three months.

5.10. For a class project, you are supposed to take a poll to forecast the outcome of an election. How many people do you have to ask so that with probability .95 your estimate will not differ from the true outcome by more than 5%?

5.11. You are planning a taste test with your brand of soft drink and another one. How large a sample would you need to be able to be 99% certain you will reject $H_0 : p = 1/2$ in favor of $H_0 : p \neq 1/2$ in a test with a type I error of 5% if only 55% of the people prefer your soft drink? To make the answer easier to see, approximate $\sqrt{0.55 \cdot 0.45}$ by $1/2$.

5.12. 24% of the residents in a community are members of a minority group but among the 96 people called for jury duty only 13 are. Does this data indicate that minorities are less likely to be called for jury duty?

5.13. A factory owner is concerned that the fraction of red M&M's produced by his machines is not $1/4$. In a sample of 1200 candies, 329 were red. Should he doubt the accuracy of his machines?

5.14. A business consultant suggested to a company that it would increase its sales if its employees wore more conservative clothes. 13 conservatively dressed salesmen sold an average of 35.6 widgets with a variance of 52 while 12 in casual dress sold an average of 28.4 widgets with a variance of 60. Are we 95% confident that there is a real difference?

5.15. Two types of plastic are being considered for use in electronic circuit boards. The breaking strength of this plastic is important so the company breaks 16 samples of type 1, finding a mean of 152.3 pounds per square inch with a standard deviation of 1.38, while 16 samples of type 2 had a mean of 151.4 and a standard deviation of 1.18. Is there a significant difference between the mean breaking strengths?

5.16. In a poll of 900 Americans in 1978, 65% said that extramarital sex was wrong, whereas a similar poll in 1985 found that 72% had the same opinion. Are we 95% confident that opinions have changed?

5.17. A basketball player made 69.2% of his 289 free throw attempts in one season and 77.2% of his 289 free throw attempts in the next one. (Yes, the data is phony, but it makes the arithmetic a little simpler.) Show that we can reject

$H_0 : p_1 = p_2$ in favor of $H_1 : p_1 \neq p_2$ with a type I error of 5% even though the 95% confidence intervals for the two seasons overlap.

5.18. A name-brand pain reliever eliminates headache pain in 70% of the cases. To test the effectiveness of our new product we set up a test of the hypothesis $H_0 : p = 0.7$ against the alternative $H_1 : p \neq 0.7$. 22 of our 25 subjects report that their headaches are cured. Show that we are not 95% confident that our product is better but a 95% confidence interval for the fraction of people who are cured by our product does not contain 0.7!

Answers to Selected Exercises

CHAPTER 1

1.1. (a) 4, (b) 3   **1.2.** $6^3 = 216$ **1.3.** (a) $2^4 = 16$, (b) 4, (c) 6

1.4. $A \cup B = \{1, 2, 3, 4\}$, $A \cap B = \{2\}$, $A^c = \{3, 4, 5, 6\}$, $B - A = \{3, 4\}$

2.1. 2,12:1/36, 3,11:2/36, 4,10:3/36, 5,9:4/36, 6,8:5/36, 7:6/36

2.4. $\Omega = \{ABC, ACB, BAC, BCA, CAB, CBA\}$, where we have represented the children by the first letters of their names.

2.6. (a) $\Omega = \{0, 1, \dots, 9\}$ with probability 1/10 each. (b) No, for two reasons: (i) within a city there at most a few dozen possibilities for the first three digits, and (ii) the first digit is never 1, which indicates a long-distance call.

2.8. (a) 4/9, (b) 5/9   **2.10.** (a) 3/4, (b) 11/16

2.12. (a) 1/216, (b) 3/216, (c) 6/216, (d) 10/216, (e) 15/216, (f) 21/216

2.14. 90   **2.16.** 0.2

3.1. $2 \cdot 11 \cdot 3 \cdot 5 = 330$ **3.2.** $2^{15} = 32,768$ **3.3.** $9! = 362,880$

3.4. $8! = 40,320$ **3.5.** $P_{11,3} = 11 \cdot 10 \cdot 9 = 990$

3.6. (a) 30^5, (b) $30 \cdot 29 \cdot 28 \cdot 27 \cdot 26$

3.7. $C_{15,4} = (15 \cdot 14 \cdot 13 \cdot 12)/(1 \cdot 2 \cdot 3 \cdot 4) = 1,365$

3.8. $C_{52,5} = (52 \cdot 51 \cdot 50 \cdot 49 \cdot 48)/(1 \cdot 2 \cdot 3 \cdot 4 \cdot 5) = 2,598,960$

3.9. (a) Take $a = b = 1$ in (3.5). (b) Take $a = -1, b = 1$.

3.10. (a) $x^5 + 10x^4 + 40x^3 + 80x^2 + 80x + 32$, (b) $8x^3 + 36x^2 + 54x + 27$

3.12. $37!/(3! 4! 5! 25!)$ **3.13.** $15!/(6! 5! 4!)$ **3.14.** $52!/(13!)^4$

3.16. (a) $26^3 10^3$, (b) $(25 \cdot 24 \cdot 9 \cdot 8)/(26^2 10^2)$ **3.18.** 86,400

3.20. (a) 1,680, (b) 3,360   **3.22.** 2/n   **3.24.** 1,152

3.26. 34,650   **3.28.** $2,520/46,656 = 0.054$ **3.30.** 1/20   **3.38.** $C_{n,m}$

3.40. $C_{m,j} C_{n-1,m-j-1}/C_{n+m-1,m-1}$ **3.42.** $C_{15,5}$

4.1. $C_{6,3} C_{47,2}/C_{54,6} \approx 1/1195$ **4.4.** 4/6

4.6. $712,842/2,598,960 = 0.2743$ **4.8.** 14/33

4.10. $P_{16,5} = 524,160$ (a) $215,040/P_{16,5} = 0.4102$, (b) $268,800/P_{16,5} = 0.5128$, (c) $40,320/P_{16,5} = 0.0767$ **4.12.** $64/1326 = 0.0482$

4.14. $210/792 = 0.2652$ **4.16.** $46,080/658,008 = 0.0700$

5.1. (a) 1/128, (b) 7/128, (c) 21/128, (d) 35/128   **5.2.** 0.1512

5.4. 120/343 = 0.3498   **5.6.** $1980/12^5 = 0.00796$

5.8. (a) 9/22 = 0.409, (b) 3/8 = 0.375

5.10. 21/32   **5.12.** (a) 0.504, (b) 0.432, (c) 0.036, (d) 0.027, (e) 0.001

6.1. Upper bound = 1.1918, lower bound = 0.4832, exact answer = 0.7063

6.2. (a) 90%, (b) 30%   **6.4.** 1/24, 6/24, 8/24, 9/24

6.6. 14,833/40,320 = 0.36788   **6.8.** 0.028

6.10. $(4 - 104/11, 951)C_{13,6}C_{39,7}/C_{52,13} = 0.1655$

6.12. 1/6, 0.1550926, 0.1555212; exact answer = 0.1555124

6.14. 1/5, 0.181, 0.18214; exact answer = 0.18209

6.18. (b) $\sum_i P(A_i) - 2\sum_{i<j} P(A_i \cap A_j) + 3\sum_{i<j<k} P(A_i \cap A_j \cap A_k)\ldots$
Erase the "..." to get the answer to (a).

7.2. (a) 151,200, (b) 210

7.4. (a) 24, (b) 576, (c) 105, (d) 2,520, (e) 40,320, (f) 384

7.6. 215,160   **7.8.** 73%   **7.10.** 5/12   **7.12.** 15/216

7.14. 0.5814   **7.16.** (a) 5/30, (b) 15/30, (c) 9/30, (d) 1/30   **7.18.** 0.42

7.20. (a) 624/783 = 0.7960, (b) 156/783 = 0.1992, (c) 3/783 = 0.0038

7.22. 10/32   **7.24.** $4 \cdot C_{13,5} \cdot 2383/(3 \cdot C_{52,7})$

CHAPTER 2

1.1. (a) Yes, (b) Yes, (c) No   **1.2.** $C_{10,3}(1/4)^3(3/4)^7$ **1.3.** $C_{10,8}(0.7)^8(0.3)^2$

1.4. $C_{8,2}(1/6)^2(5/6)^6$ **1.5.** $12(0.3)(0.1)(0.6)^2$

1.6. $(15!/10!\,3!\,2!) \cdot (0.7)^{10}(0.2)^3(0.1)^2$ **1.8.** (a) Yes, (b) No   **1.16.** 0.4

1.18. 19/27   **1.20.** (a) 1/14,400, (b) and (c): 4 1/3 times the answer for (a)

1.22. $5.48(0.6)^4 = 0.710208$ **1.24.** (a) 0.6651, (b) 0.6186   **1.26.** 13

2.2. $P(A) = 1/2, P(B) = 0.4069$ **2.3.** 0.0621   **2.6.** 7/11

2.8. (a) 1320/2197 = 0.6008, (b) 110/156 = 0.7051   **2.10.** 56/131

2.12. 7.2%   **2.14.** 1/8, 3/8, 1/2

3.6. 17%   **3.8.** 0.4   **3.10.** 0.7   **3.12.** 0.55

3.14. $146/(36)^2 = 0.1126$ **3.16.** 30/61 = 0.4918   **3.18.** Third boy, 7/18

3.20. $f_2 = 0.2, f_3 = 0.52, \lim_{k\to\infty} f_k = 3/7$ **3.22.** 3/5

4.1. 1/3, 1/9   **4.2.** 4/7   **4.6.** 0.9   **4.8.** 7/9

4.10. 1/4   **4.12.** 27/91   **4.14.** (a) 0.07, (b) 4/7, (c) 0.0928

4.16. 10/27   **4.18.** 0.8   **4.20.** 5/6

5.2. The success probabilities are (a) 0.2430, (b) 0.0799   **5.4.** ≈ 0.9

5.6. 4.39×10^{-5} **5.12.** $a_i = 1/2 + (-1/2)^i + (-1/2)(-1/3)^i$

6.2. 1/4   **6.4.** 7   **6.6.** $1/2\left(1 + (4/6)^k\right)$

6.8. (a) 0.784, (b) 0.83692   **6.10.** 0.462664

6.12. (a) 0.048, (b) 0.296, (c) 0.464, (d) 0.192   **6.14.** 5/25

6.16. (a) $146/1296 = 0.1127$, (b) $11/216 = 0.0509$ **6.18.** 0.9568

6.20. 2/3   **6.22.** 2/3   **6.24** 12/13

CHAPTER 3

1.2. 3 or 12:1/64, 4,11:3/64, 5,10:6/64, 6,9:10/64, 7,8:12/64

1.4. $P(Y = k) = (1 - e^{-\lambda})e^{-\lambda k}$ for $k \geq 0$

1.8. $1 - e^{-0.4} = 0.3296$ **1.10.** $e^{-3} = 0.0497$

1.12. $1 - 3e^{-2} = 0.5939$ **1.14.** $8.5e^{-3} = 0.4232$

2.1. (a) 0.8300, (b) 0.9772, (c) 0.2417

2.4. $6x(1 - x)$ for $0 < x < 1$, 0 otherwise

2.6. No. $F'(x) = 3 - 4x < 0$ for $3/4 < x < 1$.   **2.8.** 3/4

2.10. (a) $x^2/4$ for $0 \leq x \leq 2$, (b) 1/4, (c) 7/16

2.12. (a) $x^{1/2}$ for $0 \leq x \leq 1$, (b) 1/4

2.14. 5 is the only median

2.16. $(35/216)e^{-18} \leq 1 - \Phi(x) \leq (1/6)e^{-18} = 2.538 \times 10^{-9}$

3.4. $y^{-1/2}/2$ **3.6.** $1 - \exp(-e^x)$ **3.8.** $1/\pi(1 + x^2)$

3.10. (a) $1 - x^{-3}$, (b) $(1 - u)^{-1/3}$ **3.12.** $f(y) = (1 - y)^{-1/2}/2$

4.1. $2 \cdot \frac{2}{\pi}\left\{\int_0^{\theta_0} \frac{L}{2}\sin\theta\, d\theta + \left(\frac{\pi}{2} - \theta_0\right) \cdot \frac{1}{2}\right\}$ where $\sin\theta_0 = 1/L$ **4.2.** $(1 - r)^2$

4.4.

X	Y=0	1	2
0	6/153	24/153	15/153
1	32/153	48/153	0
2	28/153	0	0

4.6. $P(X = x, Y = y) = 2/36$ if $x < y$, $= 1/36$ if $x = y$

4.8. (a) $c = 1/48$, (b) 3/8   **4.10.** 1/10   **4.12.** 1/4   **4.14.** 1/6

4.16. $F(x, y) = \min\{x, y\}$ if $x, y > 0$ and $\min\{x, y\} \leq 1$,
    $F(x, y) = 1$ if $\min\{x, y\} > 1$, 0 otherwise

5.2. $f_X(x) = x + \frac{1}{2}$ for $0 < x < 1$, $f_Y(y) = \frac{1}{2} + 2y^3$ for $0 < y < 1$

5.4. $f_X(x) = 5x^4$ for $0 < x < 1$, $f_Y(y) = (10/3)(y - y^4)$ for $0 < y < 1$

5.6. $F_X(x) = F_{X,Y}(x, \infty) = \lim_{y \to \infty} F_{X,Y}(x, y)$ **5.8.** 7/16

5.10. $f_Y(y) = ye^{-y^2/2}$ for $y \geq 0$, 0 otherwise

5.12. (a) $F(x)^n$, (b) $1 - (1 - F(x))^n$ **6.2.** $f_{Y/X}(z) = 1/\pi(1 + z^2)$

6.4. $f(r, \theta) = rg(r)$ for $0 < r < 1$ and $-\pi < \theta < \pi$

6.10. $\frac{n(n-1)f(x)f(y)(n-2)!}{(j-1)!(k-j-1)!(n-k)!} F(x)^{j-1}(F(y) - F(x))^{k-j-1}(1 - F(y))^{n-k}$

6.12. $f_T(t) = 30e^{-3t} - 60e^{-4t} + 30e^{-5t}$

7.1. Suppose $Y = X + Z$ where $Z = \text{Poisson}(\mu - \lambda)$ is independent of X

7.6. $P(X + Y = n) = (n - 1)p^2(1 - p)^{n-2}$ for $n \geq 2$

7.8. $f_{X+Y}(z) = z/2$ for $0 < z \leq 1$, $1/2$ for $1 \leq z \leq 2$, $(3 - z)/2$ for $2 \leq z < 3$

7.12. $f_{X+Y}(z) = (\lambda\mu/(\mu - \lambda))(e^{-\lambda z} - e^{-\mu z})$ **8.2.** (a) 1/6, (b) 1/2

8.4.

Y X=0	3	6
1	.3	.15 .3
2	.1	.05 .1
	.4	.20 .4

8.6. $P(X = 0, Y = 0) = 4/9$ or $1/9$ **8.8.** Binomial$(n, \lambda/(\lambda + \mu))$

8.12. $(x/y)^{n-1}(1 - (x/y))^{m-1}(n + m - 1)!/((n - 1)!(m - 1)!y)$

8.14. $f_X(x) = 6x(1-x)$ for $0 < x < 1$, $f_{Y|X}(y|x) = 1/(1-x)$ for $0 < y < 1-x$

8.16. $f_X(x) = 3x^2/2 + x$ for $0 < x < 1$,
      $f_{Y|X}(y|x) = (3x + 4y)/(3x + 2)$ when $0 < y < 1$

9.2. 3:10/84, 2:40/84, 1:30/84, 0:4/84   **9.4.** Poisson: 0.4865, exact: 0.4914

9.6. (a) $2x - x^2$, (b) 1/4, (c) $1 - \sqrt{1/2} = 0.2928$

9.8. (a) $1 - (1 + e^x)^{-1}$, (b) 1/3   **9.10.** $c = \pi^{-1/2}e^{-1}$

9.12. $(2k + 1)/100$ **9.14.** $P(R \leq r) = r^3$

9.16. $c(1 + x^2/m)^{-(m+1)/2}$ **9.18.** $e^{-4}/2$

9.20. $f_X(x) = (3/4)(1 - x^2)$ for $-1 < x < 1$,
      $f_Y(y) = (3/2)(1 - y)^{1/2}$ for $0 < y < 1$

9.22. (a) $30x^2(1 - x)^2$ when $0 < x < 1$, (b) $2y/(1 - x)^2$ when $0 < y < 1$

9.24. 2/3

9.26. $f_T(t) = (t - 4)^2/8$ when $4 < t \leq 6$, $1/2 - (t - 6)^2/8$ when $6 \leq t < 8$

CHAPTER 4

1.2. $5 (plus his dollar back)   **1.4.** All bets have expected value $-2/38$

1.6. 0.4439   **1.8.** 2   **1.10.** 15/7   **1.12.** 4/5

1.14. $\sqrt{2/\pi}$ **1.16.** $(\rho - 1)/(\rho - 2)$ **1.18.** e

1.22. $EX = N^{-n}\{N^{n+1} - \sum_{k=1}^{N}(k-1)^n\} \approx N\{1 - \int_0^1 x^n\, dx\}$

1.24. $ER = 13.02$ **2.2.** $EX^m = (n(n+1)\cdots(n+m-1))/\lambda^m$

2.3. $EX = \lambda$, $EX^2 = \lambda^2 + \lambda$ **2.4.** At any point between B and C.

2.6. 32/10   **2.8.** $\gamma_X(z) = pz/(1 - (1-p)z)$, $EX = 1/p$

2.10. $\lambda/(\lambda^2 + 1)$ **2.16.** $(e^t - 2 + e^{-t})/t^2$

2.18. No. We would have $EX^{2n} = 0$ and hence $X = 0$ with probability 1.

2.20. 1/3   **2.22.** $5 - 2e^{-1/8} - 2e^{-1/2} - e^{-9/8}$ **2.24.** 1/6

2.30. Pick the largest m for which $25P(X \geq m) - 15$ is positive.

3.1. $EX = \mu$, $EX^2 = \mu^2 + \sigma^2$, $EX^3 = \mu^3 + 3\sigma^2\mu$

3.4. $ES_n = np$, $ES_n^2 = n(n-1)p^2 + np$

3.5. $n(n-1)\cdots(n-k+1)p^k$ if $k \leq n$, 0 if $k > n$ **3.6.** $-$$1

3.8. $-3/5$ **3.10.** $Ez^{N_k} = (pz/\{1 - (1-p)z\})^k$, $EN_k = k/p$

3.12. (a) 3.178, (b) 3.6   **3.14.** 3.114

3.16. (a) 48(1/5), (b) 39(1/14)   **3.18.** 4   **3.20.** $\sum_{j=1}^n 1/j \approx \ln n$

4.2. Mean $93/16 = 5.81$, variance $263/256 = 1.03$

4.4. Mean 3/4, variance 3/80   **4.6.** Mean 0, variance $2/\lambda^2$

4.12. $9\,\text{var}(X) + 16\,\text{var}(Y)$ **4.14.** $k(1-p)/p^2$ **4.16.** $\sum_{j=1}^n j^{-1} - j^{-2} \approx \ln n$

4.18. 67.84   **4.20.** $5p(1-p) + 20\{q - p^2\}$ where $p = (0.8)^7$ and $q = (0.6)^7$

4.22. $ER = m/(n+m)\sum_{j=1}^{n+m} j$

$\qquad \text{var}(R) = \{mn/(m+n)^2\}\sum_{j=1}^{n+m} j^2 - \{2mn/(m+n)^2(m+n-1)\}\sum_{i<j} ij$

4.24. (a) 0, (b) No

4.30. $-(nKL/N)\{1 - (n-1)/N - 1)\}$, where $N = K + L + M$

5.2. (a) 1/2, (b) $1/3 + (1/2)(X - 2/3)$

5.4. (a) $-1/11$, (b) $7/12 - (1/11)(X - 7/12)$ **5.6.** (a) $v_1/\sqrt{v_1^2 + v_2^2}$,

(b) $X + m_2$ **5.8.** (a) $\mu/(1 - \theta)$, (b) $\sigma^2/(1 - \theta^2)$, (c) θ^n

5.10. 1/2   **5.12.** (a) $-9/72$, (b) No   **5.14.** $Y = 2X + 1$

5.16. $3 \leq \text{var}(X + Y) \leq 7$ **5.18.** -1 **5.20.** $-1/(n-1)$

6.1. X is uniform on $\{-2, -1, 0, 1, 2, \}$, $Y = X^2$ **6.2.** 0.9772

6.6. 1/4   **6.8.** $EX = 15$, $EY = 5$ **6.10.** 2.8

6.12. 28   **6.14.** $E(Y|X) = 2X/3$, var$(Y|X) = X^2/18$

6.16. $E(X|Y) = Y/3$, var$(X|Y) = Y^2/18$

6.18. $P(Y = 1) = 12/36$. If $k \geq 2$, $P(Y = k) = \sum_{i=0}^{2} \frac{6+2i}{36} \frac{9+i}{36} \left(\frac{27-i}{36}\right)^{k-2}$

7.2. $(2e - 1)/(e - 1) \approx \$2.58$ **7.4.** (a) $2^{n+1} - 2$, (b) 2^n

7.6. $ER = \sqrt{\pi/2}$, var$(R) = (4 - \pi)/2$

7.8. $(n - 1)/(n + 2)$, $n(n - 1)/(n + 1)(n + 2)$, $2(n - 1)/(n + 1)^2(n + 2)$

7.10. $\theta = b/(a + b)$ **7.14.** (a) 4, (b) 0

7.16. Mean $(n + 1)/2$, variance $(n - 1)/4$ **7.18.** Mean 13/3, variance 6.14

7.20. Mean 16/7, variance $288/245 = 1.176$ **7.22.** $EA = 1/2$, var$(A) = 1/36$

7.24. -5 **7.26.** -3

CHAPTER 5

1.2. Chebyshev bound 1/4, exact probability 1/8

1.4. Chebyshev: $1/36 = 2.77 \times 10^{-2}$, Chapter 3 bound: 2.538×10^{-9}

2.2. 0.516   **2.4.** (a) 0.3085, (b) 0.2709   **2.6.** (a) 0.1336, (b) 0.0124

2.8. 0.067   **2.10.** 0.4452   **2.12.** 0.3594   **3.1.** (a) 2.33, (b) 2.58

3.2. $[11.7, 16.3]$ **3.4.** $[2.0, 2.2]$ **3.6.** $[46.5, 51.5]$ **3.8.** $[0.508, 0.528]$

3.10. $[0.3398, 0.5004]$ **3.12.** $[9/30, 11/30]$

3.14. $[312, 336]$ **3.16.** (a) $M_n/(0.05)^{1/n}$, (b) 3.091

4.2. $9/16 - 4/13 > 0.2307$, so we reject H_0

4.4. $6 < 2\sqrt{18}$, don't reject H_0 **4.6.** $6 < 2(3.33)$, don't reject H_0

4.8. $12 > 2(5)$, reject H_0 **4.10.** $3 > 2(1.24)$, reject H_0

4.12. $24 < 2(28.14)$, don't reject H_0 **4.14.** $17/150 > 2(0.05)$, reject H_0

4.16. $[-0.388, 0.908]$ **4.18.** 3/5   **5.2.** 0.2266   **5.4.** 156

5.6. (a) 0.0188, (b) 6458.4   **5.8.** (a) $[10, 12]$, (b) $[9.5, 13.5]$ **5.10.** 400

5.12. $0.1046 > 2(0.0435)$, reject H_0 **5.14.** $7.2 > 2(3)$, reject H_0

5.16. $0.07 > 2(0.02189)$, so we are quite confident opinions have changed

5.18. $0.84 < 0.8833$, don't reject H_0; confidence interval $[0.75, 1.01]$

Formulas for Important Distributions

Here (3.1.2) means that the formula was derived in Example 1.2 of Chapter 3.

Bernoulli distribution. A random variable that takes the values 1 and 0 with probabilities p and $1 - p$ is said to have a Bernoulli distribution.

$$(4.3.3) \qquad EX = p$$

$$(4.4.7) \qquad \text{var}(X) = p(1 - p)$$

Binomial distribution. If we perform an experiment n times and on each trial there is a probability p of success then the number of successes S has a binomial(n, p) distribution.

$$(3.1.2) \qquad P(S = k) = \binom{n}{k} p^k (1 - p)^{n-k} \qquad \text{for } k = 0, 1, \ldots, n$$

$$(4.3.4) \qquad ES = np$$

$$(4.4.8) \qquad \text{var}(S) = np(1 - p)$$

$$(4.3.10) \qquad Ee^{tS} = (1 - p + pe^t)^n$$

(3.7.1) If $X = $ binomial(m, p) and $Y = $ binomial(n, p) are independent then

$$X + Y = \text{binomial}(m + n, p)$$

Hypergeometric distribution. If we draw n balls out of an urn with M red and N black balls then the number of red balls R we get has

$$(3.1.1) \qquad P(R = k) = \frac{\binom{M}{k}\binom{N}{n-k}}{\binom{M+N}{n}} \qquad \text{for } k = 0, 1, \ldots, n$$

$$(4.3.6) \qquad ER = np \qquad \text{where } p = M/(M + N)$$

$$(4.4.10) \qquad \text{var}(R) = np(1 - p)\frac{T - n}{T - 1} \qquad \text{where } T = M + N$$

Geometric distribution. If we perform an experiment n times and on each trial there is a probability p of success then the number of the trial on which the first success occurs, N, has a geometric(p) distribution.

$$(3.1.3) \qquad P(N = k) = (1 - p)^{k-1} p \qquad \text{for } k = 1, 2$$

(4.1.11) $EN = 1/p$
(4.4.3) $\text{var}(N) = (1-p)/p^2$

(2.1.11) If $N_1, \ldots, N_n$ are independent geometric(p) then the sum $T_n = N_1 + \cdots + N_n$ gives the number of trials we have to wait for n successes and has the **negative binomial distribution**

$$P(T_n = n + k) = \binom{n+k-1}{k} p^n (1-p)^k$$

Poisson distribution. X is said to have a Poisson(λ) distribution if

(3.1.4) $P(X = k) = e^{-\lambda} \dfrac{\lambda^k}{k!}$ for $k = 0, 1, \ldots$

(4.1.5) $EX = \lambda$
(4.4.4) $\text{var}(X) = \lambda$
(4.2.4) $Ee^{tX} = e^{\lambda(e^t - 1)}$

(3.7.2) If $X = \text{Poisson}(\lambda)$ and $Y = \text{Poisson}(\mu)$ are independent then

$$X + Y = \text{Poisson}(\lambda + \mu)$$

Uniform distribution. X is said to have a uniform distribution on (a, b) if it has density function

(3.2.1) $f_X(x) = \begin{cases} 1/(b-a) & \text{when } a < x < b \\ 0 & \text{otherwise} \end{cases}$

(4.1.6) $EX = (a + b)/2$
(4.4.1) $\text{var}(X) = (b - a)^2/12$

Exponential distribution. X is said to have an exponential(λ) distribution if it has density function

(3.2.3) $f_X(x) = \begin{cases} \lambda e^{-\lambda x} & \text{when } x \geq 0 \\ 0 & \text{otherwise} \end{cases}$

$$(4.1.7) \qquad\qquad EX = 1/\lambda$$
$$(4.4.2) \qquad\qquad \text{var}(X) = 1/\lambda^2$$
$$(4.2.3) \qquad\qquad Ee^{tX} = \lambda/(\lambda - t)$$

The exponential distribution has the lack of memory property

$$(3.2.7) \qquad\qquad P(X > t + s | X > t) = P(X > s)$$

Gamma distribution. The sum of n independent exponential(λ) random variables has a gamma(n, λ) density function

$$(3.7.5) \qquad\qquad f_X(x) = \begin{cases} \frac{\lambda^n x^{n-1}}{(n-1)!} e^{-\lambda x} & \text{when } x \geq 0 \\ 0 & \text{otherwise} \end{cases}$$

$$(4.1.7) \qquad\qquad EX = \alpha/\lambda$$
$$(4.4.2) \qquad\qquad \text{var}(X) = \alpha/\lambda^2$$
$$(4.2.3) \qquad\qquad Ee^{tX} = \lambda^\alpha/(\lambda - t)^\alpha$$

Normal distribution. X is said to have a normal(μ, σ^2) distribution if it has density function

$$(3.3.2) \qquad\qquad f_X(x) = \frac{1}{\sqrt{2\pi\sigma^2}} e^{-(x-\mu)^2/2\sigma^2}$$

$$(4.3.1) \qquad\qquad EX = \mu$$
$$(4.4.5) \qquad\qquad \text{var}(X) = \sigma^2$$
$$(4.3.2) \qquad\qquad Ee^{tX} = \exp(\mu t + \sigma^2 t^2/2)$$

(3.7.4) If $X = \text{normal}(\mu, a)$ and $Y = \text{normal}(\nu, b)$ are independent then

$$X + Y = \text{normal}(\mu + \nu, a + b)$$

Table of the Normal Distribution

$$\Phi(x) = \int_{-\infty}^{x} \frac{1}{\sqrt{2\pi}} e^{-y^2/2}\, dy$$

To illustrate the use of the table: $\Phi(0.36) = 0.6406$, $\Phi(1.34) = 0.9099$

	0	1	2	3	4	5	6	7	8	9
0.0	0.5000	0.5040	0.5080	0.5120	0.5160	0.5199	0.5239	0.5279	0.5319	0.5359
0.1	0.5398	0.5438	0.5478	0.5517	0.5557	0.5596	0.5636	0.5675	0.5714	0.5753
0.2	0.5793	0.5832	0.5871	0.5910	0.5948	0.5987	0.6026	0.6064	0.6103	0.6141
0.3	0.6179	0.6217	0.6255	0.6293	0.6331	0.6368	0.6406	0.6443	0.6480	0.6517
0.4	0.6554	0.6591	0.6628	0.6664	0.6700	0.6736	0.6772	0.6808	0.6844	0.6879
0.5	0.6915	0.6950	0.6985	0.7019	0.7054	0.7088	0.7123	0.7157	0.7190	0.7224
0.6	0.7257	0.7291	0.7324	0.7357	0.7389	0.7422	0.7454	0.7486	0.7517	0.7549
0.7	0.7580	0.7611	0.7642	0.7673	0.7703	0.7734	0.7764	0.7793	0.7823	0.7852
0.8	0.7881	0.7910	0.7939	0.7967	0.7995	0.8023	0.8051	0.8078	0.8106	0.8133
0.9	0.8159	0.8186	0.8212	0.8238	0.8264	0.8289	0.8315	0.8340	0.8365	0.8389
1.0	0.8413	0.8438	0.8461	0.8485	0.8508	0.8531	0.8554	0.8577	0.8599	0.8621
1.1	0.8643	0.8665	0.8686	0.8708	0.8729	0.8749	0.8770	0.8790	0.8810	0.8830
1.2	0.8849	0.8869	0.8888	0.8907	0.8925	0.8943	0.8962	0.8980	0.8997	0.9015
1.3	0.9032	0.9049	0.9066	0.9082	0.9099	0.9115	0.9131	0.9147	0.9162	0.9177
1.4	0.9192	0.9207	0.9222	0.9236	0.9251	0.9265	0.9279	0.9292	0.9306	0.9319
1.5	0.9332	0.9345	0.9357	0.9370	0.9382	0.9394	0.9406	0.9418	0.9429	0.9441
1.6	0.9452	0.9463	0.9474	0.9484	0.9495	0.9505	0.9515	0.9525	0.9535	0.9545
1.7	0.9554	0.9564	0.9573	0.9582	0.9591	0.9599	0.9608	0.9616	0.9625	0.9633
1.8	0.9641	0.9649	0.9656	0.9664	0.9671	0.9678	0.9686	0.9693	0.9699	0.9706
1.9	0.9713	0.9719	0.9726	0.9732	0.9738	0.9744	0.9750	0.9756	0.9761	0.9767
2.0	0.9772	0.9778	0.9783	0.9788	0.9793	0.9798	0.9803	0.9808	0.9812	0.9817
2.1	0.9821	0.9826	0.9830	0.9834	0.9838	0.9842	0.9846	0.9850	0.9854	0.9857
2.2	0.9861	0.9864	0.9868	0.9871	0.9875	0.9878	0.9881	0.9884	0.9887	0.9890
2.3	0.9893	0.9896	0.9898	0.9901	0.9904	0.9906	0.9909	0.9911	0.9913	0.9916
2.4	0.9918	0.9920	0.9922	0.9924	0.9927	0.9929	0.9931	0.9932	0.9934	0.9936
2.5	0.9938	0.9940	0.9941	0.9943	0.9945	0.9946	0.9948	0.9949	0.9951	0.9952
2.6	0.9953	0.9955	0.9956	0.9957	0.9959	0.9960	0.9961	0.9962	0.9963	0.9964
2.7	0.9965	0.9966	0.9967	0.9968	0.9969	0.9970	0.9971	0.9972	0.9973	0.9974
2.8	0.9974	0.9975	0.9976	0.9977	0.9977	0.9978	0.9979	0.9979	0.9980	0.9981
2.9	0.9981	0.9982	0.9982	0.9983	0.9984	0.9984	0.9985	0.9985	0.9986	0.9986
3.0	0.9986	0.9987	0.9987	0.9988	0.9988	0.9989	0.9989	0.9989	0.9990	0.9990

Index